Biology of Agoraphobia

CLINICAL INSIGHTS

Biology of Agoraphobia

Edited by
JAMES C. BALLENGER, MD

Professor and Chairman,
Department of Psychiatry and Behavioral Sciences,
Medical University of South Carolina

AMERICAN PSYCHIATRIC PRESS, INC.
Washington, D.C.

Note: The authors have worked to ensure that all information in this book concerning drug dosages, schedules, and routes of administration is accurate at the time of publication and consistent with standards set by the U.S. Food and Drug Administration and the general medical community. As medical research and practice advance, however, therapeutic standards may change. For this reason and because human and mechanical errors sometimes occur, we recommend that readers follow the advice of a physician directly involved in their care or the care of a member of their family.

This monograph is based on material presented at the 136th Annual Meeting of the American Psychiatric Association. That meeting and this monograph are endeavors to share scientific findings and new ideas. The opinions expressed in this monograph are those of the individual authors and not necessarily those of the American Psychiatric Association.

Library of Congress Cataloging in Publication Data

Main entry under title:
Biology of agoraphobia.

(Clinical insights)
"Based on the five papers prepared for the Symposium 'Neurobiology of Panic Disorder and Agoraphobia,' presented at the May 1983 meeting of the American Psychiatric Association in New York"—Introd.
1. Agoraphobia—Physiological aspects—Congresses.
2. Anxiety—Physiological aspects—Congresses. 3. Panic—Physiological aspects—Congresses. I. Ballenger, James C.
II. American Psychiatric Association. III. Symposium "Neurobiology of Panic Disorder and Agoraphobia" (1983: American Psychiatric Association) IV. Series.
[DNLM: 1. Phobic disorders—Etiology—Congresses.
2. Phobic disorders—Physiopathology—Congresses.
3. Fear—Physiology—Congresses. WM 178 N494 1983]
RC535.B56 1984 616.85'225 84-6157
ISBN 0-88048-064-5 (pbk.)

Printed in the U.S.A.

Contents

Contributors

JAMES C. BALLENGER, MD
Professor and Chairman,
Department of Psychiatry and Behavioral Sciences,
Medical University of South Carolina

JEAN-PHILIPPE BOULENGER, MD
Visiting Associate,
Unit on Anxiety and Affective Disorders,
Biological Psychiatry Branch,
National Institute of Mental Health; and
French National Institute of Health and Medical Research

DANIEL B. CARR, MD
Instructor in Medicine,
Harvard Medical School; and
Clinical and Research Fellow in Medicine,
Department of Medicine—Endocrine Division,
Massachusetts General Hospital

DANIEL J. COX, PhD
Associate Professor,
Department of Behavioral Medicine and Psychiatry,
University of Virginia Medical Center

ABBY J. FYER, MD
Assistant Professor of Clinical Psychiatry and
Co-Director, Anxiety Disorders Clinic,
New York State Psychiatric Institute and
Columbia Presbyterian Medical Center

JACK M. GORMAN, MD
Assistant Professor of Clinical Psychiatry and
Director, Biological Studies Unit,
New York State Psychiatric Institute and
Columbia Presbyterian Medical Center

William R. Hobbs, MD
Associate Professor,
Department of Behavioral Medicine and Psychiatry,
University of Virginia Medical Center

Andrew Hucek, MD
Clinical Assistant Professor,
Department of Behavioral Medicine and Psychiatry,
University of Virginia Medical Center

David Jimerson, MD
Acting Chief, Section on Experimental Therapeutics,
Laboratory of Clinical Science,
National Institute of Mental Health

Donald F. Klein, MD
Director, Psychiatric Research,
New York State Psychiatric Institute; and
Professor of Psychiatry,
College of Physicians and Surgeons,
Columbia University

C. R. Lake, MD, PhD
Professor of Psychiatry and Pharmacology,
Uniformed Services University of the Health Sciences
Hebert School of Medicine

Michele Laraia, RN, MSN
Clinical Research Coordinator,
Department of Psychiatry and Behavioral Sciences,
Medical University of South Carolina

Andrew P. Levin, MD
Research Fellow, Anxiety Disorders Clinic,
New York State Psychiatric Institute and
Columbia Presbyterian Medical Center

Michael R. Liebowitz, MD
Assistant Professor of Clinical Psychiatry and
Director, Anxiety Disorders Clinic,
New York State Psychiatric Institute and
Columbia Presbyterian Medical Center

Gregory A. Peterson, MD
Fellow in Forensic Psychiatry,
Department of Behavioral Medicine and Psychiatry,
University of Virginia Medical Center

Robert M. Post, MD
Chief, Biological Psychiatry Branch,
National Institute of Mental Health

David V. Sheehan, MB, BCh, BAO
Assistant Professor of Psychiatry,
Massachusetts General Hospital,
Harvard Medical School

James R. Shipe, Jr., PhD
Analytical Chemist,
Department of Pathology,
University of Virginia School of Medicine

Carol Trockman, RN
Program Nurse Specialist II,
Department of Psychiatry and Behavioral Sciences,
Medical University of South Carolina

Thomas W. Uhde, MD
Chief, Unit of Anxiety and Affective Disorders,
Biological Psychiatry Branch,
National Institute of Mental Health

Bernard J. Vittone, MD
Medical Staff Fellow,
Unit on Anxiety and Affective Disorders,
Biological Psychiatry Branch,
National Institute of Mental Health

Clinton Wilkinson, PhD
Department of Behavioral Medicine and Psychiatry,
University of Virginia Medical Center

Introduction

This monograph is based on the five papers prepared for the symposium "Neurobiology of Panic Disorder and Agoraphobia," presented at the May 1983 meeting of the American Psychiatric Association in New York. The symposium featured reports of ongoing biologically oriented research in agoraphobia and panic disorder from four academic centers (Virginia, Columbia, the National Institute of Mental Health, and Massachusetts General Hospital). The papers presented were all preliminary reports of work in progress. The papers as presented here reflect further work in these laboratories since May 1983, but each of the studies remains unfinished and in progress. Complete reports from these investigative groups will appear when this work is further developed; this is especially true of Chapters 1, 2, and 5. The sample sizes reported in the series of studies in Chapter 1 are small and are being expanded. Certain data are as yet unavailable from the work reported in Chapter 2. Chapter 5 presents an integrated hypothesis of recent and historical findings in the biology of panic disorder and is from the group at the Massachusetts General Hospital. This hypothesis is currently under experimental investigation, and the data related to this model of panic anxiety are evolving.

Chapter 1 is from the National Institute of Mental Health group headed by Dr. Thomas Uhde from the Unit on Anxiety and Affective Disorders in the Biological Psychiatry Branch, which is directed by Dr. Robert M. Post. Others who have contributed to the work and manuscript include Larry J. Siever, M.D., Marilla Geraci, R.N., Stephen Sinclair, and Patricia W. Hafer.

Dr. Uhde reviews the historical evolution of the concepts of anxiety, especially as they relate to biological conceptions of anxiety. This should serve as an introduction to the remainder of the monograph, as well as to introduce perhaps the most intriguing biological model of panic anxiety—that is, the locus coeruleus model proposed by Dr. Eugene Redmond and colleagues at Yale University. In this model, locus coeruleus activity in primates is associated with the fear/arousal/alarm reaction. As Drs. Uhde (Chapter 1) and Ballenger (Chapter 2) review, locus coeruleus (LC) activity is increased in fear-inducing situations, and medications which reduce LC firing are anxiolytic. Conversely, medications which increase LC firing are anxiogenic.

Dr. Uhde reviews the data relating this model to the sudden paroxysmal anxiety observed in panic anxiety and reports studies from his laboratory and others of the effect of clonidine on panic anxiety. Clonidine's antipanic activity is thought to be related to its action of reducing LC firing via a presynaptic alpha-2 inhibitory receptor. He also reviews the potent anxiogenic action of yohimbine, which is thought to produce anxiety by increasing LC firing rates.

In Chapter 2, Dr. James C. Ballenger of the Medical University of South Carolina in Charleston reviews the clinical studies of this model of anxiety in agoraphobic patients. This work was primarily performed in the Department of Behavioral Medicine and Psychiatry at the University of Virginia Medical School where Dr. Ballenger was Director of Research. Others contributing substantially to the study and the preparation of the manuscript include Ginny Sprouse, Vicki Brumbelow, and Pamela Thompson.

This chapter reviews evidence that the norepinephrine and epinephrine systems in agoraphobic patients appear to be "overactive" even between episodes of panic anxiety. This is certainly consistent with the locus coeruleus model and is among the early data indicating a presumably central role of the LC and noradrenergic systems in panic anxiety. The tricyclic and monoamine oxidase inhibitor antidepressants, which have both been shown to be effective in treating panic anxiety, decrease LC firing. This has been postulated to be their mechanism of action in this condition. Data reviewed in Chapter 2 of the effect of imipramine, a tricyclic antidepressant, on the pattern of the peripheral measures of total norepinephrine and epinephrine activity are certainly consistent with their action being to reduce LC firing and "whole body" (nor)adrenergic activity. This chapter also reviews the relationship between serum levels of antidepressants and the clinical response in agoraphobics. This chapter reports the largest systematic study to date in this area of investigation.

Dr. William Hobbs, senior author of Chapter 3, is also from the Department of Behavioral Medicine and Psychiatry at the University of Virginia in Charlottesville. Others contributing to the work and manuscript include Ann Thompson and Hilary Hoge.

Dr. Hobbs reviews the area of the psychophysiology of agoraphobia and reports ongoing studies of psychophysiologic patterns in agoraphobic patients which suggest that there may be a specific psychophysiologic response pattern in these patients. He also identifies pretreatment psychophysiologic measures which are associated with positive responses to treatment with imipramine. Additionally, he reports the first investigation of the intriguing accounts by many agoraphobic patients that they experience more panic attacks under fluorescent lighting. In fact, agoraphobic patients when studied in his laboratory do report increases in anxiety under fluorescent lighting in a controlled, counterbalanced research design.

In Chapter 4, Dr. Levin reports the ongoing studies of the biology of the lactate induction of panic. This review was supported in part by NIMH grants NH 33422 and NH 30906 and by research science development award MH 00416-51 (Dr. Gorman).

Dr. Levin reviews the fascinating history of the discovery and exploration of sodium lactate's ability to stimulate a typical panic attack in agoraphobic patients. It is now well established that the infusion of sodium lactate produces a typical panic attack in approximately 75–85 percent of these patients. Explanations for this phenomenon have been offered by multiple investigators, but the New York Psychiatric Institute group has performed the most in-depth investigations into the biology of the lactate-induced panic attack. Chapter 4 reviews the potential mechanisms of action for lactate's induction of panic in comprehensive detail. Although the work of this group remains in progress and as yet offers no conclusive explanation, understanding of this phenomenon perhaps offers our best current opportunity for a more basic understanding of the biology of this condition. The reproducibility of this phenomenon in the laboratory setting, and the apparent potential to understand lactate's biologic actions, make this area of investigation a very promising, albeit difficult, area of investigation.

In Chapter 5, Dr. Daniel Carr, from the Endocrine Division of the Department of Medicine, and Dr. David Sheehan of the Department of Psychiatry at the Massachusetts General Hospital, report one aspect of their investigative activities. Although this group has been extremely active in the studies of psychopharmacological treatment of this disorder, this chapter deals with their theoretical attempts to integrate various recent and historical findings of the pathophysiology of this disorder. Their attempt to explain certain prominent pulmonary symptoms and the metabolic and clinical effects of lactate and carbon dioxide (CO_2) via a common mechanism involving the midbrain center sensitive to changes in pH, bicarbonate, and CO_2 is an intriguing one. This model is under active experimental study at this point, but provides another potential biological model of panic anxiety. As the authors are quick to point out, the chemoreceptor zone in the midbrain does influence activity of the locus coeruleus and that these two biological models of anxiety may not be contradictory.

Partial support for this research was provided by grant RR 1066 from the National Institutes of Health and a Dalance fellowship of the American Philosophical Society (Dr. Carr). Dr. Carr was also supported by a grant-in-aid from the Upjohn Company.

Although much of the work reported in this monograph is preliminary, the authors hope that this monograph provides background for the clinician in this increasingly important clinical syndrome which affects 1–6 percent of the population. Since it represents a significant proportion of the work being done in the biology of this condition at this point in time, we hope that it will also serve as a review for workers in this field.

James C. Ballenger, M.D.

1

Historical and Modern Concepts of Anxiety: A Focus on Adrenergic Function

Thomas W. Uhde, M.D.
Jean-Philippe Boulenger, M.D.
Bernard J. Vittone, M.D.
Robert M. Post, M.D.

1

Historical and Modern Concepts of Anxiety: A Focus on Adrenergic Function

Anxiety is a subjective emotion ubiquitous in the human species. From a functional perspective, the experience of anxiety most commonly serves to "alarm" the biological system so that appropriate behavior for the task at hand can be initiated, directed, and sustained. Under extreme conditions this may take the form of flight or fight. In more typical circumstances, anxiety may arouse and sustain those biological functions required to meet the challenges of daily living (e.g., taking an examination, meeting a deadline, avoiding an accident). This concept emphasizes the positive *alarm* and *organizing* functions of anxiety, a perception too often forgotten by "phobophobic" patients and clinicians alike.

Yet anxiety can have as its principal consequence disorganization and disruption of goal-directed behavior. In these circumstances, anxiety becomes "pathological." Traditional attempts to classify anxiety have employed a combination of two basic approaches, descriptive and quantitative. In the former case, one emphasizes the qualitative nature of anxiety (free-floating, anticipatory, situational, phobic), while the quantitative scheme conceptualizes anxiety on a continuum from normal to extreme (terror or panic). Whether these types of anxiety reflect similar or disparate underlying biological substrates (or whether biological systems are even relevant to the topic of anxiety) has been debated for decades and remains, even today, a controversial issue among investigators in this area.

EARLY NEUROBIOLOGICAL THEORIES

In the late nineteenth century, James (1890) focused on the biological aspects of human emotion, particularly fear and rage. James assumed a direct linkage between peripheral bodily changes and human emotion. In fact, he viewed alterations in bodily sensations *as* the emotion. He even introduced the concept that some individuals might be unusually sensitive to their own internal physiological cues, a concept that has recently been suggested as a partial model to explain the apparent vulnerability to "spontaneous" and chemically induced panic attacks in agoraphobic patients. James's hypothesis stimulated much interest in the autonomic nervous system and the neurobiology of emotion. In the early twentieth century, Cannon (1928, 1929) demonstrated that emotional stimuli may induce adrenal gland secretion. However, he subsequently reported that although the exogenous administration of epinephrine produced palpitations and chest discomfort there was no simultaneous experience of fear, as would have been predicted by the James–Lange theory of human emotion.

These and other findings led Cannon to shift his attention to the study of the brain and related systems that might be involved in the neurobiology of fear, anxiety, arousal, and other human emotions. On the basis of a series of studies in the cat, Cannon eventually proposed that thalamic discharges simultaneously produced both an emotion, including fear, and alterations in the autonomic and peripheral nervous systems. Thus, James and Cannon as well as other investigators in the late nineteenth and early twentieth centuries (for review, see Plutchik 1980) addressed what today remains an unresolved issue regarding the nature and relative importance of the peripheral versus the central nervous systems (CNS) in the neurobiology of human anxiety. Like these early investigators, much of modern science has directed its attention toward the study of the adrenergic nervous system in an attempt to improve our understanding of the important biological functions of fear and anxiety.

ADRENERGIC FUNCTION

In modern psychiatry, both the peripheral and central adrenergic systems have been the focus of study in relation to the neurobiology of anxiety, fear, and arousal. Epinephrine, produced primarily by the adrenal glands in the peripheral nervous system, acts mostly as a beta-adrenergic agonist and is more influenced by stress than norepinephrine (NE), which acts primarily at the alpha-adrenergic receptor (Dimsdale and Moss 1980; Akerstedt et al. 1983; Kopin 1981).

The Peripheral Adrenergic System

Increased peripheral adrenergic activity has been demonstrated in some (Chosy et al. 1970) but not all (Morrow and Labrum 1978) anxious patients. Decreased adrenergic indices have been reported following successful treatment with relaxation techniques (Hoffman et al. 1982; Mathew et al. 1980). Stress (Akerstedt et al. 1983) and panic attacks induced by flooding techniques (Edmonson et al. 1972) or lactate (Appleby et al. 1981) are associated with increases in plasma epinephrine without significant changes in NE. The possible influence of plasma epinephrine on platelet monoamine oxidase (MAO) activity (Gentil et al. 1976; Owen et al. 1977) may also be an explanation for the increased platelet MAO activity found in patients with anxiety disorders (Kralik et al. 1983; Yu et al. 1983) and for the decreased activity associated with the anti-anxiety effects of various treatments (Kralik et al. 1983; Davidson et al. 1981; Sullivan et al. 1978; Mathew et al. 1981). In addition, plasma concentrations of free fatty acids (FFA), which are regulated in part by beta-adrenergic receptors, may be increased experimentally by epinephrine administration (Pinter et al. 1967) or in various anxiety-provoking situations (Brown and Heninger 1975; Gittleman et al. 1968; Gottschalk et al. 1965). Moreover, epinephrine elevates the plasma concentration of lactate, an effect that may underlie the increased lactate concentration reported in anxiety disorders (Jones and Mellersh 1946; Linko 1950; Cohen and White 1950).

Clinical and electrophysiological studies also suggest that the peripheral symptoms of anxiety are primarily related to activation of the sympathetic adrenal system, the source of most peripheral epinephrine (for review, see Lader 1976; Kelly 1980). Furthermore, epinephrine has been shown in many studies to be more effective than NE in inducing both peripheral and central anxiety-like symptoms (for review see Guttmacher et al. 1983). Epinephrine's beta-adrenergic effects compared to NE suggest the involvement of this system in the peripheral expression of anxiety symptoms, a hypothesis that is further substantiated by the clinical effects of agonists and antagonists of beta-adrenergic receptors.

In humans, numerous studies have shown that beta blockers are able to attenuate the somatic (and to a lesser extent the psychological) symptoms of anxiety (for review, see Pitts and Allen 1979; Noyes et al. 1981). Though other beta blockers that do not cross the blood-brain barrier, such as practolol, may also improve the psychological symptoms of anxiety (Bonn et al. 1972), the involvement of a central mechanism in the anxiolytic activity of beta blockers that may be able to reach the CNS, such as propranolol, is still a matter of controversy (Gottschalk et al. 1974; Taylor et al. 1981; Kathol et al. 1980). In addition, some but not all investigators have reported that propranolol prevents the occurrence of spontaneous panic attacks (Heiser 1976; Kathol et al. 1980) but not that of lactate-induced panic attacks (Gorman et al. 1983). Further studies are obviously required to substantiate these preliminary reports, which suggest an interesting hypothetical chain of events (i.e., epinephrine stimulates beta-adrenergic receptors that in turn mediate lactate secretion which triggers a panic attack).

The effects of beta-adrenergic receptor agonists also suggest that this system may be involved in the expression of various anxiety symptoms. Beta-adrenergic stimulants are able to increase anxiety symptoms in stressful situations and to induce physiological modifications similar to those characteristic of anxiety (Erdmann and Van Linden 1980; Brantigan et al. 1982). Furthermore, patients with *hyperdynamic beta-adrenergic state*, a condition clinically related to anxiety disorder, display an increased sensitivity to isoproterenol, a beta-adrenergic agonist. These patients have "emotional outbursts" following the administration of isoproterenol, which are prevented by propranolol (Frohlich et al. 1969). This condition may be close to that of the patients described by Easton and Sherman (1976). They found that sufferers of somatic anxiety had attacks precipitated by isoproterenol infusions at rates that would not induce such symptoms in normals (Cleaveland et al. 1972). These patients also improved on propranolol. A comparable finding has been described in patients suffering from other syndromes similar to anxiety disorders (Hughes et al. 1981). It is also of interest that patients with mitral valve prolapse, a condition frequently associated with panic disorders (Kantor et al. 1980), may be hyper-responsive to isoproterenol (Boudoulas et al. 1980).

The involvement of beta-adrenergic receptors in the pathogenesis of anxiety disorders is further suggested by the fact that a reduction in the number of brain beta-adrenergic receptors may be one of the biochemical factors underlying adaptation to stress (Stone and Platt 1982). This effect is also induced by chronic tricyclic or MAO antidepressant therapy (for review, see Charney et al. 1981), a treatment that is useful in panic disorder, where patients seem to be especially sensitive to stress (Uhde et al. 1984a). To our knowledge, the binding of beta-receptor ligands on human lymphocytes has never been studied in anxiety disorders, but some studies strongly suggest a hypersensitivity of these receptors in various clinical entities whose symptomatology is close, if not identical, to that of primary anxiety disorders (Hughes et al. 1981; Easton and Sherman 1976; Frohlich 1969). It is noteworthy that in hyperthyroidism, in which anxiety and panic attacks are part of the clinical picture, an increase in the number of beta-adrenergic receptors in various tissues has been reported both in animals and humans (for review, see Motulsky and Insel 1982).

Although these adrenergic indices are most closely correlated with peripheral adrenergic function, other lines of evidence suggest that the central noradrenergic system may be activated as well in relation to various anxiety states.

The Central Adrenergic System

Cerebrospinal fluid studies have demonstrated a positive correlation between concentrations of NE and state-anxiety measures in depressed patients (Post et al. 1980), and between measures of anxiety in normal volunteers and concentrations of 3-methoxy-4-hydroxyphenyl glycol (MHPG), the major metabolite of NE (Ballenger et al. 1984). It has been estimated that up to 60 percent of the peripheral pool of MHPG originates from central origins (Maas et al. 1979). Furthermore, several lines of data suggest that central and peripheral noradrenergic systems may be linked to each other and to the hypothalamo–pituitary–adrenal (HPA) axis (Crawley et al. 1978; Elsworth et al. 1982; Jimerson et al. 1981, 1983; Sotsky et al. 1981) in a number of physiological states. Thus, it is noteworthy that plasma MHPG and state-anxiety are positively correlated in normal volunteers (Uhde et al. 1982a) and panic disorder patients (Ko et al. 1983). Also, urinary MHPG concentrations have been reported to covary with changes in anxiety in affectively ill patients (Sweeney et al. 1978). Together, these data are consistent with several lines of animal and human research suggesting a role for the central NE system in stress (Kopin 1981) and anxiety.

THE LOCUS COERULEUS

Most of the data presented in the preceding sections are consistent with an animal model of anxiety developed by Redmond and colleagues at Yale University. His investigations provide evidence that the noradrenergic nucleus locus coeruleus (LC), a nucleus situated in the pons with projections to the cerebral cortex, cerebellum, limbic system, brain stem, reticular system, and the pain-sensitive neurons in the dorsal horn of the spinal cord, may be part of an anatomic substrate for the biological functions of alarm, anxiety, fear, and arousal (Redmond 1977, 1979; Redmond and Huang 1979). Redmond demonstrated that electrical or pharmacologic activation of the LC produced behaviors in the stump-tailed monkey similar to those observed following exposure to natural threat in the wild. Furthermore, pharmacologic activation of the locus coeruleus in the monkey is associated with increased levels of NE and MHPG in various areas of the brain and in the cerebrospinal fluid and plasma (Redmond 1977, 1979; Redmond and Huang 1979), changes which parallel the association between anxiety and noradrenergic indices reported in humans. While some aspects of Redmond's model have been questioned (Mason and Fibiger 1979; File et al. 1979; Koob et al. 1980), there are some interesting parallels between the effects of drugs on LC activity in various animal preparations and their properties in humans. Although exceptions clearly exist, agents that increase noradrenergic function tend to have anxiogenic effects in humans.

In contrast, pharmacologic agents that decrease LC activity generally have anxiolytic profiles. Thus, Korf and colleagues (Korf et al. 1974) and Bird and Kuhar (1977) reported that systemically and iontophoretically applied morphine reduced spontaneous activity of LC neurons, although tolerance to these acute inhibitory effects appear to occur with chronic administration (Aghajanian 1978). This acute inhibition, followed by tolerance in the animal, is of interest in relation to observations in man that while opiates initially may have potent anxiolytic properties, this effect may wear off with chronic administration. Ethanol, an agent frequently abused to self-medicate dysphoria, has been reported to decrease (Pohorecky and Brick 1977; Strahlendorf and Strahlendorf 1981b) or have no effect (Aston-Jones et al. 1982) on spontaneous LC activity. Benzodiazepines (Redmond 1977; Grant et al. 1980; Strahlendorf and Strahlendorf 1981a) and the antipanic tricyclic antidepressants inhibit LC firing (Nyback et al. 1975; Svensson and Usdin 1978). Propranolol (Redmond 1979) blocks the postsynaptic effects of noradrenergic activity.

Since clonidine, an alpha-2 agonist, is also a potent inhibitor of LC activity in animal preparations (Aghajanian 1978), we and other investigators tested the hypothesis that clonidine might have antianxiety properties in humans. As an additional probe of the adrenergic system, our unit (Anxiety and Affective Disorders) also initiated studies with yohimbine, an alpha-2 antagonist that increases LC activity.

SELECTED DRUGS WITH ADRENERGIC EFFECTS

This section reviews the literature and presents new findings from our laboratory suggesting that clonidine (an alpha-2 agonist) and yohimbine (an alpha-2 antagonist) have anxiolytic and anxiogenic properties, respectively, in humans.

Clonidine

Gold et al. (1978) and Uhde et al. (1980) demonstrated that clonidine reversed and suppressed the emergence of most signs and symptoms of opiate withdrawal, a condition associated with profound apprehension in humans and increased LC firing in animals.

Svensson et al. (1978) reported in an uncontrolled trial that clonidine had antianxiety properties in six of 13 hospitalized patients with severe anxiety. Liebowitz et al. (Liebowitz et al. 1981; Liebowitz, personal communication) obtained antipanic effects in nine of 11 patients with open clonidine, although several patients developed tolerance with chronic administration. Hoehn-Saric et al. (1981) found clonidine (administered blindly) superior to placebo in the treatment of patients with generalized anxiety as well as panic disorder. This latter finding is noteworthy given the interesting but inconclusive evidence suggesting that imipramine is effective in the treatment of panic but not for generalized anxiety. These findings are also intriguing in light of the generally accepted observations that benzodiazepines given at the usual clinical doses effective in the management of generalized anxiety have little or no effect in the treatment of spontaneous panic attacks. The fact that alprazolam, a new and potent triazolobenzodiazepine, appears to block panic attacks indicates that potency issues may have contributed to the earlier negative studies with benzodiazepines in the treatment of panic disorder (McNair and Kahn 1981). It is also likely, however, that alprazolam shares some properties with tricyclic antidepressants. An emerging body of evidence thus suggests that clonidine may have antianxiety properties across a wide spectrum of conditions, including the opiate-withdrawal syndrome and major affective, generalized anxiety and panic disorders. Eventually it will be important to compare clonidine's antianxiety properties to those of standard anxiolytics. In the only double-blind crossover study to address this question, Ko et al. (1983) found clonidine and imipramine equally effective in reducing anxiety in patients exposed to phobic stimuli that had previously triggered panic attacks.

As part of an ongoing anxiety research and treatment program at the National Institute of Mental Health (NIMH), we have assessed the acute antianxiety properties of clonidine in anxious depressed and panic-anxious patients (Uhde et al. 1981, 1982b, 1984b; Siever et al. 1982; Siever and Uhde, 1984). Intravenous clonidine, 2μg/kg, was given to 18 anxious depressed patients, 11 agoraphobic patients with panic attacks, and 19 normal controls to assess the effect of this drug on state-anxiety. The clonidine challenges in the depressed patients were performed in collaboration with Dr. Larry Siever (Uhde et al. 1981, 1982b, 1984b; Siever and Uhde 1984; Siever et al. 1981, 1982). As illustrated in Figure 1, clonidine had robust antianxiety effects as indicated by changes on the Spielberger State Anxiety Scale (Spielberger et al. 1970) in both patient groups ($p<.006$), although both the placebo response and anxiolytic effects of clonidine in the patients with panic disorder appeared to be more robust. Euthymic normal controls had no antianxiety response to clonidine, a finding also reported by Leckman et al. (1980), but which may be related to low baseline ratings of anxiety. In fact, we have reported previously that clonidine-induced changes in measures of pain insensitivity in normal volunteers might be related to a reduction in anxiety, suggesting that clonidine might have anxiolytic properties in euthymic individuals temporarily exposed to anxiety provoking situations. Although the panic patients tended to improve following placebo, clonidine's placebo-corrected effects remained highly significant ($p = .01$). There also was no relationship between the degree of placebo response and clonidine's anxiolytic effects in the panic patients. Although clonidine produced noteworthy drowsiness, there was no correlation between the anxiolytic and sedative properties of clonidine. In the combined group of depressed patients and normal controls, those subjects with the highest baseline levels of plasma MHPG had the most robust decrement in measures of anxiety after clonidine ($r = .41$, $p < .03$). It will be of interest to learn in our upcoming investigations whether adrenergic indices in the panic disorder group also predict clinical response to clonidine.

In an ongoing placebo-controlled clinical trial, we also have found clonidine to be effective, particularly during the initial period of drug administration. In a subgroup of patients, however, tolerance to clonidine's anxiolytic effects develops after several

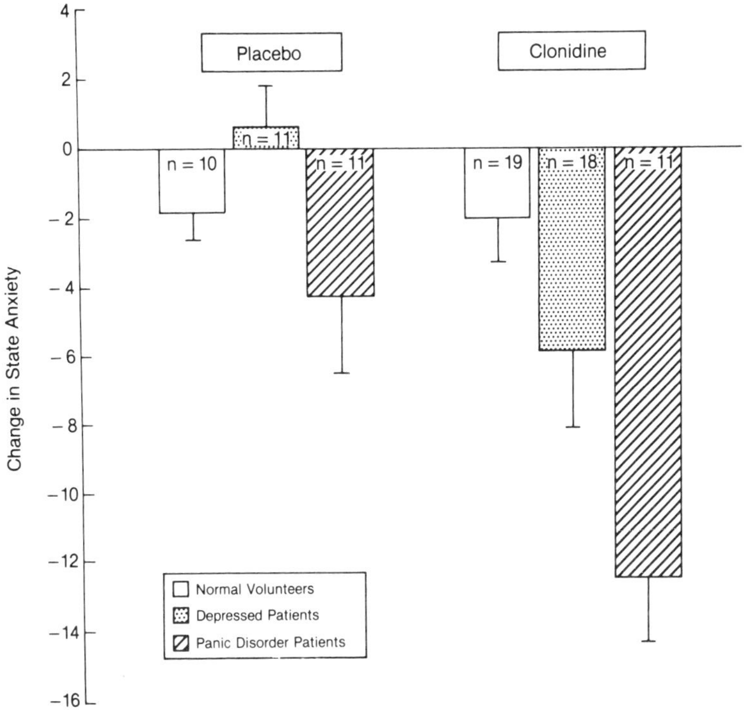

Figure 1. Acute anxiolytic effects of clonidine in depressed and panic-anxious patients. Clonidine produced significant antianxiety effects as measured by a reduction in state-anxiety on the Spielberger Anxiety Scale in anxious depressed and panic-anxious patients but not normal controls. Placebo had no significant effect in any group, although panic-anxious patients tended to improve for several hours after the administration of placebo.

weeks (Uhde et al. 1982b). This clinical observation was of particular interest in relation to Redmond's hypothesis (1977) when it was subsequently demonstrated by Svensson (personal communication) that tolerance may rapidly develop to the ability of clonidine to decrease firing of the LC. From a clinical perspective, it should be further underscored that clonidine may be associated with a number of untoward side effects including daytime drowsiness and lethargy, depression, dry mouth, and hypotension. Thus, because of these potential side effects and its nonapproved status by the Food and Drug Administration for the treatment of psychiatric illnesses, we would not recommend clonidine as a drug of first choice in the treatment of panic and related anxiety disorders. However, in treatment of refractory panic disorder, a clinical trial of clonidine might represent a reasonable alternative. Furthermore, clonidine, a well-known and effective antihypertensive agent, might be useful in those hypertensive patients with prominent symptoms of anxiety.

Yohimbine

Since yohimbine (an alpha-2 antagonist) has effects on LC firing opposite those of clonidine (an alpha-2 agonist), yohimbine would be expected to have anxiogenic properties in humans. Furthermore, since yohimbine has relatively selective effects on adrenergic function, it might represent a useful research tool for investigating adrenergic substrates of human anxiety and the neurobiological mechanisms of different antianxiety treatments.

Consistent with the hypothesis that noradrenergic overactivity may be involved in some states of anxiety, yohimbine has been reported by several investigators to produce anxiety-like symptoms in humans (Garfield et al. 1967; Holmberg and Gershon 1961; Charney et al. 1982, 1983), especially when administered in high doses intravenously (Garfield et al. 1967; Holmberg and Gershon 1961; and Ingram 1962). To confirm these observations and to study the effects of yohimbine in a population at risk for panic attacks, we administered low-dose oral yohimbine (20 mg) to agoraphobic patients with spontaneous panic attacks. Six of the first seven patients studied to date had profound anxiety after

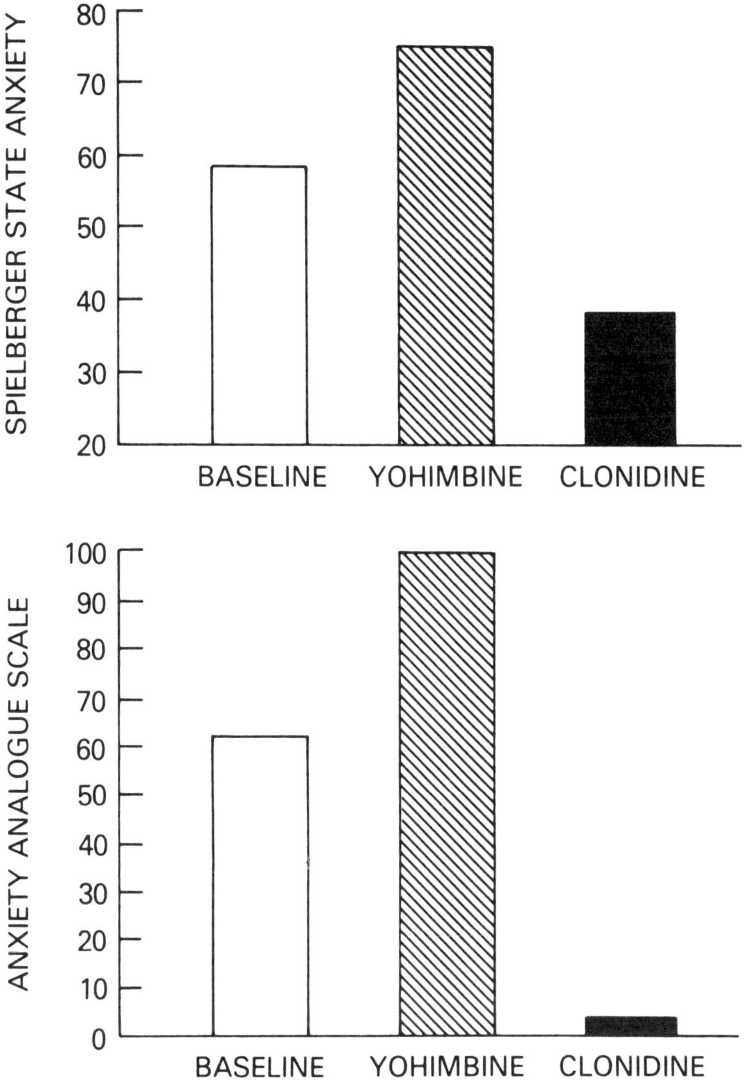

Figure 2. Opposite effects of yohimbine and clonidine in a panic-anxious patient. In this patient with panic disorder, yohimbine, an alpha-2 antagonist, produced profound anxiety as indicated by an increase to near maximal scores on two separate measures of self-rated anxiety. Clonidine, an alpha-2 agonist, not only immediately reversed the yohimbine-induced anxiety but also decreased measures of anxiety far below levels obtained at baseline.

yohimbine, including five patients who had panic attacks, while none had anxiety to this degree or panic attacks while receiving placebo as part of other challenge paradigms ($p < .05$). As illustrated by the patient in Figure 2, yohimbine (20 mg) produced essentially maximal increases on two separate measures of anxiety, which were completely reversed by a single oral dose of clonidine. In fact, clonidine reduced measures of anxiety in this patient to levels well below those obtained at baseline. This apparent increased vulnerability to low-dose yohimbine ($\leq$20 mg) in panic disorder patients is not typical of normal subjects (Charney et al. 1982), reflecting perhaps an alteration in noradrenergic function or receptor sensitivity in panic disorder. It should be underscored, however, that the anxiogenic properties of yohimbine might be mediated through benzodiazepine receptors, dopaminergic, or other neurotransmitter systems (Scatton et al. 1980; Marwaha and Aghajanian 1982; Lal et al. 1983), although the most reasonable explanation at this point is one of increased or overactive adrenergic function. It is noteworthy in this regard that the two patients in our study who had the most severe anxiety after a yohimbine challenge were subsequently our two best responders to a clinical trial with clonidine.

Charney et al. (1983) found in normal subjects that pretreatment with either clonidine (5 μg/kg) or diazepam (10 mg) antagonized yohimbine-induced (30 mg) anxiety, while clonidine but not diazepam diminished yohimbine-induced increases in plasma MHPG. These data suggest that while adrenergic overactivity may be important in some forms of anxiety, low-moderate doses of diazepam probably mediate their principal antianxiety effects through a nonadrenergic neurotransmitter system. It is interesting, however, that diazepam at this dose also does not block spontaneous panic attacks. It would be of considerable interest, therefore, to determine whether alprazolam, a benzodiazepine which apparently blocks panic attacks, attenuates both yohimbine-induced anxiety and plasma MHPG. Such paradigms will be useful in clarifying the biological similarities and differences among different anxiety disorders and mechanisms by which anxiolytic agents mediate their therapeutic effects.

CONCLUSIONS

The investigations of James and Cannon explored the relationships among human emotion, peripheral and central anatomical structures, and neurophysiological function. Recent data, rather than clarifying the contribution of central and peripheral mechanisms, have obscured the issue, since peripheral and central adrenergic systems appear to change in parallel in response to many environmental and pharmacological manipulations. Experimental clinical trials of drugs that do not cross the blood-brain barrier may be helpful in this regard. However, few investigators today would take the position that alterations in adrenergic function, or any other neurotransmitter system, are exclusively relevant to the topic of anxiety. Indeed, the frame of reference might productively shift to more discrete research strategies and questions. For instance, in which patients are adrenergic alterations important and to which aspects of the patients' acute symptomatology and longitudinal course are these changes related? What is the role of "nonspecific" stress and more specific psychologically relevant precipitants in the panic anxiety syndromes; are some aspects related to adrenergic function dependent on emotional context and conditioning?

Thus, adrenergic mechanisms may mediate important aspects of normal arousal and pathological anxiety in some patients, but often may involve physiological mechanisms involving benzodiazepines, gamma aminobutyric acid (GABA), serotonin, or other neurotransmitter systems. Although it is possible that many of these systems could impinge on and modulate relatively discrete neuroanatomical areas involved in anxiety, few would argue that the search for a single "anxiety center" is likely to be successful. Although central–peripheral distinctions cannot always be clearly delineated, the accumulation of evidence over the past several hundred years suggests an important role for adrenergic function in the neurobiology of fear, anxiety, and arousal.

Given the above caveats, it would nonetheless appear that investigations of neurochemical, physiological, and behavioral responses to agents with relatively selective mechanisms of action (e.g., clonidine and yohimbine) will increase our understanding of the role of adrenergic and other neurotransmitter systems in both normal and pathological anxiety. This strategy (Siever et al. 1981) may also help elucidate mechanisms by which drugs mediate their therapeutic effects and lead to the development of new anxiolytic agents.

Already, based on Redmond's research on locus coeruleus function in the stump-tailed monkey, clinical studies in humans have been initiated and clonidine has been found to reduce anxiety in a number of psychiatric disorders. In this fashion the animal model based on locus coeruleus function and the associated noradrenergic hypothesis of anxiety have already proved heuristically valuable. One is reminded of the fate of a similar catecholaminergic hypothesis of manic depressive illness, which has both generated vast amounts of data but has not yet been unequivocally validated or disproven. However, the ability to predict biochemical–behavioral relationships, develop new drugs, and formulate more specific research questions and models to be proven or disproven may justify our current primitive attempts at model building and testing. Ultimately, all animal "models," including the noradrenergic hypothesis based on locus coeruleus function, will be judged on their usefulness in accurately predicting whether drugs will have anxiolytic or anxiogenic effects in humans.

References

Aghajanian GK: Tolerance of locus coeruleus neurons to morphine and suppression of withdrawal response by clonidine. Nature 276:186–187, 1978

Akerstedt T, Gillberg M, Hjemdahl P, et al: Comparison of urinary and plasma catecholamine responses to mental stress. Acta Physiol Scand 117:19–26, 1983

Appleby IL, Klein DF, Sachar EJ, et al: Biochemical indices of lactate-induced panic: a preliminary report, in Anxiety: New Research and Changing Concepts. Edited by Klein DF, Rabkin J. New York, Raven Press, 1981

Aston-Jones G, Foote GSL, Bloom FE: Low doses of ethanol disrupt sensory responses of brain noradrenergic neurones. Nature 196:857–860, 1982

Ballenger JC, Post RM, Jimerson DC, et al: Neurobiological correlates of depression and anxiety in normal individuals, in Neurobiology of Mood Disorders. Edited by Post RM, Ballenger JC. Baltimore, Williams & Wilkins, 1984, pp 481–501

Bird SJ, Kuhar MJ: Iontophoretic application of opiates to the locus coeruleus. Brain Res 122:523–533, 1977

Bonn JA, Turner P, Hicks DC: Beta-adrenergic receptor blockade with practolol in the treatment of anxiety. Lancet 1:814–815, 1972

Boudoulas H, Wooley CF, Reynolds J: Mitral valve prolapse syndrome: hyperresponsiveness to adrenergic stimulation. Clinical Research 28:755, 1980

Brantigan CO, Brantigan TA, Joseph N: Effect of beta-blockade and beta stimulation on stage fright. Am J Med 72:88–94, 1982

Brown WA, Heninger G: Cortisol, growth hormone, free fatty acids and experimentally evoked affective arousal. Am J Psychiatry 132:1172–1176, 1975

Cannon WB: Neural organization for emotional expression, in Feelings and Emotions: The Wittenberg Symposium. Edited by Regmert ML. Worcester, Mass, Clark University Press, 1928

Cannon WB: Bodily Changes in Pain, Hunger, Fear and Rage. New York, Appleton, 1929

Charney DS, Menkes DB, Heninger GR: Receptor sensitivity and the mechanism of action of antidepressant treatment. Arch Gen Psychiatry 38:1160–1180, 1981

Charney DS, Heninger GR, Sternberg DE: Assessment of alpha-2 adrenergic autoreceptor function in humans: effects of oral yohimbine. Life Sci 30:2033, 1982

Charney DS, Heninger GR, Redmond DE Jr: Yohimbine induced anxiety and increased noradrenergic function in humans: effects of diazepam and clonidine. Life Sci 33:19–29, 1983

Chosy JJ, Lewis WC, Graham DT, et al: Phobia questionnaire responses and urine catecholamines. Arch Gen Psychiatry 22:58–62, 1970

Cleaveland CR, Rangno RE, Shand DG: A standardized isoproterenol test. Arch Intern Med 130:47–52, 1972

Cohen ME, White PD: Life situations, emotions and neurocirculatory asthenia (anxiety neurosis, neuresthenia, effort syndrome). Proceedings of The Association for Research in Nervous and Mental Disease 29:832–869, 1950

Crawley JN, Hattox SE, Maas JW, et al: 3-Methoxy-4-hydroxyphenethylene glycol increase in plasma after stimulation of the locus coeruleus. Brain Res 141:380–384, 1978

Davidson J, Linnoila M, Raft D, et al: MAO inhibition and control of anxiety following amitriptyline therapy. Acta Psychiatr Scand 63:147–152, 1981

Dimsdale JE, Moss J: Plasma catecholamines in stress and exercise. JAMA 243:340–342, 1980

Easton JD, Sherman DG: Somatic anxiety attacks and propranolol. Arch Neurol 33:689–691, 1976

Edmonson HD, Roscoe B, Vickers MD: Biochemical evidence of anxiety in dental patients. Br Med J 4:7–9, 1972

Elsworth JD, Redmond DE Jr, Roth RH: Plasma and cerebrospinal fluid 3-methoxy-4-hydroxyphenylethylene glycol (MHPG) as indices of brain norepinephrine metabolism in primates. Brain Res 235:115–124, 1982

Erdmann G, Van Linden B: The effects of beta-adrenergic stimulation and beta-adrenergic blockade on emotional reactions. Psychophysiology 17:331–338, 1980

File SE, Deakin JFW, Longden A, et al: An investigation of the role of the locus coeruleus in anxiety and agonistic behavior. Brain Res 169:411–420, 1979

Frohlich ED, Tarazi RC, Dustan HP: Hyperdynamic beta-adrenergic circulatory state. Increased beta-receptor responsiveness. Am J Med 123:1–7, 1969

Garfield S, Gershon S, Sletten I, et al: Chemically induced anxiety. International Journal of Neuropsychiatry 3:426–433, 1967

Gentil V, Greenwood MH, Lader MH: The effect of adrenaline on human platelet MAO activity. Psychopharmacologia 44:187–190, 1976

Gittleman B, Shatin L, Birenbaum M, et al: Effects of quantified stressful stimuli on blood lipids in man. J Nerv Ment Dis 147:196–201, 1968

Gold MS, Redmond DE Jr, Kleber HD: Clonidine in opiate withdrawal. Lancet 1:929, 1978

Gorman JM, Levy GF, Liebowitz MR, et al: Effect of acute beta-adrenergic blockade on lactate-induced panic. Arch Gen Psychiatry 40:1079–1089, 1983

Gottschalk LA, Cleghorn JM, Gleser GC: Studies of relationships of emotions to plasma lipids. Psychosom Med 27:102–111, 1965

Gottschalk LA, Stone WN, Gleser GC: Peripheral versus central mechanisms accounting for antianxiety effects of propranolol. Psychosom Med 36:47–56, 1974

Grant SJ, Huang YH, Redmond DE Jr: Benzodiazepines attenuate single unit activity in the locus coeruleus. Life Sci 27:2231–2236, 1980

Guttmacher LB, Murphy DL, Insel TR: Pharmacologic models of anxiety. Compr Psychiatry 24:312–326, 1983

Heiser JF, Defrancisco D: The treatment of pathologic panic states with propranolol. Am J Psychiatry 133:1389–1394, 1976

Hoehn-Saric R, Merchant AF, Keyser ML, et al: Effects of clonidine on anxiety disorder. Arch Gen Psychiatry 38:1278–1282, 1981

Hoffman JW, Benson H, Arns PA, et al: Reduced sympathetic nervous system responsivity associated with the relaxation response. Science 215:190–192, 1982

Holmberg G, Gershon S: Autonomic and psychic effects of yohimbine hydrochloride. Psychopharmacologia 2:93-106, 1961

Hughes FC, Munera Y, Dellas JA, et al: Hypersensibilite des recepteurs beta-adrenergiques chez des patients classes comme spasmophiles: resultats de deux epreunes de stimulation par l'isoprenaline chez 105 patients. Ann Med Interne (Paris) 132:390–393, 1981

Ingram CG: Some pharmacologic actions of yohimbine and chlorpromazine in man. Clin Pharmacol Ther 3:345–352, 1962

James W: The Principles of Psychology. New York, Holt, Rinehart & Winston, 1890

Jimerson DC, Insel TR, Reus VI, et al: Effects of psychotropic agents on high anxiety subjects. Psychopharmacol Bull 17:97–100, 1981

Jimerson DC, Insel TR, Reus VI, et al: Increased plasma MHPG in dexamethasone-resistant depressed patients. Arch Gen Psychiatry 40:173–176, 1983

Jones M, Mellersh V: A comparison of the exercise response in anxiety states and normal controls. Psychosom Med 8:180–187, 1946

Kantor JS, Zitrin CM, Zeldis SM: Mitral valve prolapse syndrome in agoraphobic patients. Am J Psychiatry 37:467–469, 1980

Kathol RG, Noyes R Jr, Slymen DJ, et al: Propranolol in chronic anxiety disorders. A controlled study. Arch Gen Psychiatry 37:1361–1365, 1980

Kelly D: Anxiety and Emotions. Physiological Basis and Treatment. Springfield, Ill., Charles C Thomas, 1980

Ko GN, Elsworth JD, Roth RH, et al: Panic-induced elevation of plasma MHPG levels in phobic-anxious patients. Arch Gen Psychiatry 40:425–430, 1983

Koob GF, Strecker RE, Roberts DCS, et al: Failure to alter anxiety or the anxiolytic properties of chlordiazepoxide and ethanol by destruction of the dorsal noradrenergic system. Society for Neuroscience Abstracts 6:108, Abstract #40.6, 1980

Kopin IJ: Catecholamines, adrenal hormones, and stress, in Neuroendocrinology. Edited by Krieger DT, Hughes JC. Sunderland, Sinauer Association, 1981

Korf J, Bunney BS, Aghajanian GK: Noradrenergic neurons: morphine inhibition of spontaneous activity. Eur J Pharmacol 25:165–169, 1974

Kralik PM, Ho BT, Mathew RJ, et al: Kinetic evaluation of platelet monoamine oxidase following relaxation in chronic anxiety. Acta Psychiatr Scand 67:307–314, 1983

Lader MH: Anxiety, in The Psychophysiology of Mental Illness. London, Routledge & Kegan Paul, 1976

Lal H, Shearman G, Bennett D, et al: Yohimbine: beta-carboline with behavioral and neurochemical properties common to anxiogenic disorders. Society for Neuroscience Abstracts 9:437, Abstract #128.11, 1983

Leckman JF, Maas JW, Redmond DE Jr: Effects of oral clonidine on plasma 3-methoxy-4-hydroxy-phenylethyleneglycol (MHPG) in man: preliminary report. Life Sci 26:2179–2185, 1980

Liebowitz MR, Fyer AJ, McGrath P, et al: Clonidine treatment of panic disorder. Psychopharmacol Bull 17:122–123, 1981

Linko E: Lactic acid response to muscular exercise in neurocirculatory asthenia. Ann Med Int Fin 36:161–176, 1950

Maas JW, Hattox SE, Greene NM, et al: 3-Methoxy-4-hydroxy-phenethyleneglycol production by human brain in vivo. Science 205:1025, 1979

Marwaha J, Aghajanian G: Relative potencies of alpha-1 and alpha-2 antagonists in the locus coeruleus, dorsal rophe and dorsal lateral geniculate nuclei: an electrophysiological study. J Pharmacol Exp Ther 222:287–293, 1982

Mason ST, Fibiger HC: Current concepts. I. Anxiety: the locus coeruleus disconnection. Life Sci 25:2141–2147, 1979

Mathew RJ, Ho BT, Kralik P, et al: Catechol-o-methyltransferase and catecholamines in anxiety and relaxation. Psychiatry Res 3:85–91, 1980

Mathew RJ, Ho BT, Kralik P, et al: Catecholamines and monoamine oxidase activity in anxiety. Acta Psychiatr Scand 63:245–252, 1981

McNair DM, Kahn RJ: Imipramine compared with a benzodiazepine for agoraphobia, in Anxiety: New Research and Changing Concepts. Edited by Klein DF, Rabkin J. New York, Raven Press, 1981

Morrow GR, Labrum AH: The relationship between psychological and physiological measures of anxiety. Psychol Medicine 8:95–101, 1978

Motulsky HJ, Insel PA: Adrenergic receptors in man. Direct identification, physiologic regulation and clinical alterations. N Engl J Med 307:18–29,1982

Noyes R Jr, Kathol R, Clancy J, et al: Antianxiety effects of propranolol: a review of clinical studies, in Anxiety: New Research and Changing Concepts. Edited by Klein DF, Rabkin JG. New York, Raven Press, 1981

Nyback HV, Walters JR, Aghajanian GK, et al: Tricyclic antidepressants: effects on firing rate of brain noradrenergic neurons. Eur J Pharmacol 32:302–312, 1975

Owen F, Acker W, Bourne RC, et al: The effect on human monoamine oxidase activity of subcutaneous injections of adrenalin. Biochem Pharmacol 26:2065–2067, 1977

Pinter EJ, Peterfy G, Cleghorn JM, et al: The influence of emotional stress of fat metabolism: the role of endogenous catecholamines and the beta-adrenergic receptors. Am J Med Sci 254:634, 1967

Pitts FN, Allen RE: Biochemical induction of anxiety, in Phenomenology and Treatment of Anxiety. Edited by Fann WE, Karachan I, Porkorny AD, et al. New York, Spectrum Publications, 1979

Plutchik R: Emotion: A Psychoevolutionary Synthesis. New York, Harper & Row, 1980

Pohorecky LA, Brick J: Activity of neurons in the locus coeruleus of the rat: inhibition by ethanol. Brain Res 131:174–179, 1977

Post RM, Ballenger JC, Goodwin FK: Cerebrospinal fluid studies of neurotransmitter function in manic and depressive illness, in Neurobiology of Cerebrospinal Fluid. Edited by Wood JH. New York, Plenum Press, 1980

Redmond DE Jr: Alterations in the function of the nucleus locus coeruleus: a possible model for studies of anxiety, in Animal Models in Psychiatry and Neurology. Edited by Hanin I, Usdin E. Oxford/New York, Pergamon Press, 1977

Redmond DE Jr: New and old evidence for the involvement of a brain norepinephrine system in anxiety, in Phenomenology and Treatment of Anxiety. Edited by Fann WE, Karachan I, Porkorny AD, et al. New York, Spectrum Publications, 1979

Redmond DE Jr, Huang YH: Current concepts. II. New evidence for a locus coeruleus–norepinephrine connection with anxiety. Life Sci 25:2149–2162, 1979

Scatton B, Zivkovic B, Dedek J: Antidopaminergic properties of yohimbine. J Pharmacol Exp Ther 215:494–499, 1980

Siever LJ, Uhde TW: New studies and perspectives on the noradrenergic receptor system in depression: effects of the alpha-2-adrenergic agonist clonidine. Biol Psychiatry 19:131, 1984

Siever LJ, Insel TR, Uhde TW: Noradrenergic challenges in the affective disorders. J Clin Psychopharmacol 1:192–206, 1981

Siever LJ, Uhde TW, Silberman EK, et al: The growth hormone response to clonidine as a probe of noradrenergic receptor responsiveness in affective disorder patients and controls. Psychiatry Res 6:171–183, 1982

Sotsky SM, Lake CR, Goodwin FK: Adrenocortical function and plasma norepinephrine in normal human subjects. Biol Psychiatry 16:643–651, 1981

Spielberger CB, Gorsuch RL, Luschene RE: STAI Manual for the State–Trait Anxiety Inventory. Palo Alto, Calif, Consulting Psychologist Press, Inc, 1970

Stone EA, Platt JE: Brain adrenergic receptors and resistance to stress. Brain Res 237:405–414, 1982

Strahlendorf HK, Strahlendorf JC: Iontophoretically applied benzodiazepines inhibit locus coeruleus unit activity. Society for Neuroscience Abstracts 7:793, Abstract #256.2, 1981a

Strahlendorf JC, Strahlendorf HK: Response of locus coeruleus neurons to direct application of ethanol. Society for Neuroscience Abstracts 7:312, Abstract #104.7, 1981b

Sullivan JL, Dackis C, Stanfield CN: Clinical correlates of tricyclic antidepressant mediated inhibition of platelet monoamine oxidase. Biol Psychiatry 13:399–407, 1978

Svensson T, Persson R, Wallin L, et al: Anxiolytic action of clonidine. Nordisk Psykiatrisk Tidsskrift 32:439–441, 1978

Svensson TH, Usdin T: Feedback inhibition of brain noradrenaline neurons by tricyclic antidepressants: alpha-receptor mediation. Science 202:1089–1091, 1978

Sweeney DR, Maas JW, Heninger GR: State-anxiety, physical activity, and urinary 3-methoxy-4-hydroxyphenethylene glycol excretion. Arch Gen Psychiatry 35:1418–1423, 1978

Taylor EA, Jefferson D, Carroll JD, et al: Cerebrospinal fluid concentrations of propranolol, pindolol and acetolol in man: evidence for central actions of beta-adrenoreceptors antagonists. Br J Clin Pharmacol 12:549–559, 1981

Uhde TW, Redmond DE Jr, Kleber JH: Clonidine suppresses the opioid abstinence syndrome without clonidine-withdrawal symptoms: A blind inpatient study. Psychiatry Res 2:37–47, 1980

Uhde TW, Post RM, Siever LJ, et al: Clonidine: effects on mood, anxiety, and pain. Psychopharmacol Bull 17:125–126, 1981

Uhde TW, Siever LJ, Post RM, et al: The relationship of plasma-free MHPG to anxiety and psychophysical pain in normal volunteers. Psychopharmacol Bull 18:129–132, 1982a

Uhde TW, Boulenger JP, Siever LJ, et al: Animal models of anxiety: implications for research in humans. Psychopharmacol Bull 18:47–52, 1982b

Uhde TW, Boulenger JP, Roy-Byrne PP, et al: Longitudinal course of panic disorder: clinical and biological considerations. Prog Neuropsychopharmacol (in press), 1984a

Uhde TW, Siever LJ, Post RM: Clonidine: acute challenge and clinical trial paradigms for the investigation and treatment of anxiety disorders, affective illness, and pain syndromes, in Neurobiology of the Mood Disorders. Edited by Post RM, Ballenger JC. Baltimore, Williams & Wilkins, 1984b

Yu PH, Bowen RC, Davis BA, et al: Platelet monoamine oxidase activity and trace acid levels in plasma of agoraphobic patients. Acta Psychiatr Scand 67:188–194, 1983

2

A Study of Plasma Catecholamines in Agoraphobia and the Relationship of Serum Tricyclic Levels to Treatment Response

James C. Ballenger, M.D.
Gregory A. Peterson, M.D.
Michele Laraia, R.N., M.S.N.
Andrew Hucek, M.D.
C. R. Lake, M.D., Ph.D.
David Jimerson, M.D.
Daniel J. Cox, Ph.D.
Carol Trockman, R.N.
James R. Shipe, Jr., Ph.D.
Clinton Wilkinson, Ph.D.

2

A Study of Plasma Catecholamines in Agoraphobia and the Relationship of Serum Tricyclic Levels to Treatment Response

This chapter addresses two main areas. The first part deals with the clinical response of agoraphobic patients to imipramine and forms part of the growing literature showing that tricyclic antidepressants (TCAs) are effective in this condition (Klein 1964, 1967; Klein and Fink 1962; Zitrin et al. 1978, 1980, 1983; Sheehan et al. 1980; McNair and Kahn 1981; Marks 1983; Mavissakalian et al. 1984). To our knowledge, this is the second controlled investigation designed specifically to investigate the relationship between clinical treatment response and the plasma level of TCAs.

In the second part of the chapter, we examine the relationship of the plasma catecholamines, norepinephrine (NE) and epinephrine (E), to agoraphobia and its symptoms, and their relationship to the response to imipramine. As discussed in Chapter 1 and below, several lines of evidence suggest a link between the central NE system and anxiety and—perhaps in particular—panic and phobic anxiety.

Until very recently, there has been little attention given to plasma levels of the TCAs in the studies of TCAs and agoraphobia. Most studies utilize the same range of oral doses of the TCAs in agoraphobia as in depression: 150–300 mg per day. There is considerable clinical experience that some patients who do not respond at the lower doses do respond favorably subsequently when the dose is increased, suggesting that a linear response would be expected between serum level and clinical response. However, Zitrin and colleagues (Zitrin et al. 1978) have reported that 18 percent of their patients develop "stimulatory" (amphetamine-like) side effects at low doses (5–75 mg per day) of imipramine. These side effects have often resulted in dropouts or the patient's refusal to attempt higher doses. Some patients never are able to tolerate more than 5–10 mg per day. Curiously, many of these same patients report complete blockade of their panic attacks and remission of other symptomatology on these low doses. To explore whether these very small doses were actually reaching standard serum levels, perhaps because of lower metabolic rates, Jobson and colleagues measured serum TCA levels in two agoraphobic patients on 25 mg per day of imipramine and another on 25 mg per day of amitriptyline (Jobson et al. 1978). The two imipramine patients achieved combined imipramine and desipramine (its principal metabolite) levels of 55 ng/ml and 15 ng/ml. The amitriptyline patient had similarly low levels, 5–10 ng/ml.

In one of the few studies reporting no overall therapeutic effects for imipramaine in agoraphobics, plasma levels were examined (Marks et al. 1983). These patients received a mean oral dose of 158 mg per day at week 14 and 110 mg per day at week 26. Combined imipramine and desipramine levels reached mean levels of 220 ng/ml by weeks 10 through 14. Outcome at 14, 28, and 52 weeks had no relationship to plasma levels at week 14. Mavissakalian and colleagues very recently have reported a study more specifically designed to explore these relationships (Mavissakalian et al. 1984). In 15 subjects, oral doses of imipramine reached approximately 2.5 mg/kg per day, with a mean ± SD dose of 153.3 ± 28.14 mg per day (range, 125–200 mg per day). Mean plasma concentrations attained were 142 ± 116.5 ng/ml for imipramine, 100 ± 58.1 ng/ml for desipramine, and 241 ± 161.4 ng/ml for total (imipramine plus desipramine). Plasma imipramine correlated significantly with improvement, with higher levels of imipramine being associated with lower symptom scores after 12 weeks of treatment. Levels of desipramine did not correlate significantly with improvement, nor did total levels provide further information than imipramine levels alone.

In one study with obsessive–compulsive patients, who may share some characteristics with agoraphobics, plasma levels of chlorimipramine and desmethylchlorimipramine were examined. Overall, patients responded best in the middle range of plasma concentrations, similar to the inverted-U curve (*therapeutic window*) observed with nortriptyline in depressed patients. However, there are no data or even a suggestion of a therapeutic window in the published studies with agoraphobic patients.

If there is an optimal serum level of imipramine for this condition, or a minimal level, it would be important to know this to facilitate care of these patients. Most agoraphobics are quite reluctant to take any medications for fear that the medicines will worsen, not benefit, their illness. Also, as mentioned, 15 to 25 percent actually have significant "stimulatory" side effects at low doses, and the physician is uncertain how to proceed. If both patient and clinician knew that a certain serum level was required for adequate response, patient compliance in attaining that level would be maximized. Also, side effects make agoraphobic patients feel they are getting worse. Because these side effects seem related both to the rate of increase of medication and to serum level, the use of serum levels could allow the optimal amount of medication to be utilized and the avoidance of unnecessarily high doses. Obviously, if a therapeutic window were present it would have important clinical implications as well.

Multiple lines of evidence have suggested involvement of the NE and E systems in the panic anxiety observed in agoraphobia. The symptoms of the hyperdynamic beta-adrenergic circulatory state (Frohlich et al. 1969; Easton and Sherman 1976) are very similar to panic attacks with palpitations, chest discomfort, and tachycardia after exercise. The transient episodes of anxiety seen in this syndrome, reproduced by infusions of the beta-adrenergic agonist isoproterenol, also can be blocked by maintenance treatment with the beta blocker, propranolol (Easton et al. 1976). The symptomatic mitral valve prolapse patients who also have so many symptoms similar to agoraphobic patients have been shown to have high levels of circulating NE and E (Boudoulas et al. 1980). These levels also correlate with the anxiety symptomatology, and the authors concluded that the symptomatology of mitral valve patients is secondary to a high adrenergic tone. Agoraphobic patients have also been reported to have a higher than expected incidence of mitral valve prolapse (Kantor et al. 1980). However, the relationship of mitral valve prolapse to agoraphobia remains unclear.

In many experimental strategies in laboratory animals, stress has been shown to be associated with an increase in the turnover of NE (for review, see Stone 1975). These have included foot shock (Thierry et al. 1968), exposure to strangers (Bliss and Ailion 1969), and cold exposure (Gordon et al. 1966).

In humans, peripheral NE measures have been shown to be increased in chronically anxious patients (Mathew 1980) and in normal individuals in anxiety-provoking situations. These have included pilots landing on aircraft carriers (Rubin et al. 1970), parachutists (Bloom 1963), and students during exams (Jones et al. 1968). Dental phobics have increased plasma NE when in the dental chair (Edmonson et al. 1972). In two studies of depressed patients, anxiety was seen to correlate with plasma NE and E (Sweeney et al. 1978; Wyatt et al. 1971). Cerebrospinal fluid (CSF) NE was shown to correlate with anxiety in depressed patients (Post et al. 1978). In the only study of the relationship of CSF noradrenergic measures and anxiety in normals, CSF 3-methoxy-4-hydroxy-phenylethylene glycol (MHPG), 3,4-dehydroxy-phenylethylene glycol (DHPG) (CNS metabolites of NE), and dopamine beta-hydroxylase (DBH) were significantly correlated with transient, brief *state* increases in anxiety (Ballenger et al. 1984). In the plasma, there were significant negative relationships observed between *trait* anxiety and plasma MHPG and DBH. Recent studies have documented that with panic attacks induced by an infusion of sodium lactate, plasma NE increases, but E remains unchanged (see Chapter 4). Those patients who panic also had higher baseline NE levels before lactate infusion.

As reviewed in Chapter 1 of this volume, Redmond and colleagues at Yale have proposed that the midbrain noradrenergic nucleus, the locus coeruleus (LC), may be involved in primates in the brain response to alarm, fear, or anxiety (Redmond and Huang

1971). This nucleus has 70 percent of the brain noradrenergic neurons, with widespread projections to limbic and cortical areas. Low-intensity electrical stimulation of the LC results in behaviors observed when the animal is threatened (Bertrand 1969; Redmond et al. 1976a). Bilateral lesions of the LC decrease these "fearful" behaviors (Redmond et al. 1976b; Huang, et al. 1976).

Alpha adrenergic agonists like clonidine decrease LC firing as well as these "fearful" behaviors (Gold et al. 1977). Conversely, alpha adrenergic blockers like yohimbine or piperoxane increase LC activation and these behaviors (Redmond 1977; Redmond 1979; Huang et al. 1977). Piperoxane, when given to humans, resulted in severe anxiety (Holmberg et al. 1961).

A direct pharmacological test of this model is provided by the reports of clonidine's efficacy, at least transiently, in blocking panic anxiety. As mentioned, clonidine decreases LC firing by a presynaptic alpha-adrenergic inhibitory receptor.

Firing of the LC also results in brain, CSF, and plasma increases in NE and its principal metabolite MHPG. Although the exact percentage of plasma MHPG coming from the CNS is unknown, it is generally felt to reflect the activity of the LC and the total sympathetic nervous system. The LC and peripheral sympathetic neurons are linked and generally act in concert (Elsworth et al. 1982; Crawley et al. 1978; Jimerson et al. 1981a; DeMet 1979). A recent study reported increases in plasma MHPG when agoraphobic patients had panic attacks in their phobic situations (Ko et al. 1983). Also the clonidine-induced decreases in MHPG correlated with decreases in anxiety symptomatology. Tricyclic antidepressants have been shown to decrease LC firing (Nyback et al. 1975), and, in the study of Ko and colleagues, imipramine did suppress the increase in MHPG levels with panic. The monoamine oxidase (MAO) inhibitors also decrease LC firing and are effective in the treatment of agoraphobia (Sheehan et al. 1980). Nies and Robinson (1981) have shown that treatment of anxious, depressed patients with the MAO inhibitor antidepressant phenelzine results in significant decreases in plasma NE and its metabolite DHPG, but not in E. Again, these decreases paralleled the decreases in anxiety ($R = 0.55$).

At the inception of this study, it was our hypothesis that agoraphobic patients have a propensity to panic attacks because of high pre-existing NE and E "tone." We predicted they would have higher circulating levels of NE and E than normals, a prediction that was confirmed. We further predicted that the high NE and E levels would correlate with the severity of the syndrome; however, that prediction was not supported by our data. We hypothesized that the TCAs were effective in this condition because they decreased firing of the LC and, therefore, their effect should be to decrease such a postulated abnormally high adrenergic tone. We predicted we would be able to observe this effect by decreases in plasma NE, E, and MHPG, and we did observe significant decreases in E and MHPG, but increases in NE. This chapter reports data currently available from this study and further data will be reported at a later date.

METHODS

Subjects

The sample consisted of 36 patients meeting the *Diagnostic and Statistical Manual of Mental Disorders*, third edition (DSM-III), criteria for agoraphobia. All were between the ages of 18 and 65 years, had no serious physical illnesses, and had never been psychotic.

Procedures

Patients were treated with imipramine but randomly assigned to one of two treatment groups: one was maintained at 100–150 ng/ml serum level of imipramine plus its principal metabolite desipramine, and the other at 200–250 ng/ml. Serum levels were measured by high-performance liquid chromatography (HPLC) every one to two weeks throughout the study period of three months, and doses were adjusted to maintain levels in the assigned serum range. Both the patient and the investigator rating the patients were blind to which serum level the patient had been assigned. Identical capsules of 25 and 50 mg imipramine and placebo were utilized to maintain the blind.

All patients attended weekly group psychotherapy for 12 weeks in groups of five to 10, where they were educated about agoraphobia, supported, encouraged, and asked to develop a hierarchy of their feared situations. They were asked to gradually reapproach their phobic situations (generally starting with the least feared situation) after they had been on medication three weeks, and were praised for their progress in this exposure "homework." They were told we hoped the medication would block their panic attacks, and that they would need to relearn that they could re-enter their phobic situation safely without having a panic attack by actually re-exposing themselves to those situations on many occasions.

All patients received imipramine at one of the two serum levels and behaviorally oriented group psychotherapy. That is, this study did not employ a placebo group, nor was there a cell that did not receive group and exposure homework.

Rating Scales

The treating and rating psychiatrist filled out the following scales at pretreatment, six, and 12 weeks: (1) Symptom Severity and Avoidance Behavior Scale (Table 1). This is a five-point scale adapted from Marks (1969), which rates the severity of the principal phobic symptomatology and the resultant avoidance behavior. This has been widely utilized in studying agoraphobics (Sheehan et al. 1980). (2) Work and Social Disability Scale (Table 2). A second five-point clinical scale, the Work and Social Disability Scale, is also frequently used in studying agoraphobics (Kelly et al. 1970; Marks 1969; Marks et al. 1966; Sheehan et al. 1980). This

Table 1 Symptom Severity and Avoidance Behavior Scale

Score	Definition
1	No uneasiness when meeting feared object or situation
2	Uneasiness but no avoidance of feared object or situation
3	Definite fear, tendency to avoid feared object or situation
4	Strong fear, avoidance if at all possible of the feared object or situation
5	Terrifying panic attack when avoidance of feared object or situation impossible

Table 2 Work and Social Disability Scale

Score	Definition
5	Symptoms radically change or prevent normal work or social activities
4	Symptoms interfere with normal work or social activities markedly but they are not prevented or radically changed
3	Symptoms interfere with normal work or social activities in minor ways
2	Symptoms mild, but not interfering with normal work or social activities
1	No complaints, normal activity

scale and the previous one estimate the actual amount of clinical improvement based on external behavioral guidelines. To examine response on these two major outcome scales, decreases on these scales were categorized as marked, improved, no change, and worse. For the Work and Social Disability Scale, decreases of four, three, or two were considered *marked* (see Table 2). A drop of one was labeled *improved*, and any increase in ratings as *worse*. For the Symptom Severity and Avoidance Behavior Scale any improvement of four or three, or a drop of two if the final score was one or two, was considered *marked*. The other categories were the same as with the other scale. (3) The Hamilton Rating Scale for Depression (HAM-D) was scored at pretreatment (Hamilton 1960).

Patients filled out the following scales at pretreatment, six, and 12 weeks.

(1) The Symptom Checklist-90 (SCL-90) (Derogatis et al. 1974)
(2) The Wolpe–Lange Fear Survey (Wolpe and Lange, 1964)
(3) The Beck Depression Inventory (BDI) (Beck and Beamesderfer 1974)
(4) Half of the patients filled out scales measuring global improvement, improvement in the number of panic attacks, and intensity of avoidance of their main phobias. These data will be reported elsewhere.

Biological Measures

At pretreatment, six, and 12 weeks, the following biological measures were obtained:

(1) Plasma samples for norepinephrine (NE) and epinephrine (E) were obtained via an indwelling catheter after 30 minutes of supine bed rest (NE-S, E-S) and after five minutes of standing. The change with standing is thought to represent the reactivity of the sympathetic nervous system, including the baroreceptor reflex arc through the vasomotor regulatory centers in the brain and the outflow to the peripheral sympathetic nerve endings and adrenal medulla. Change is reported as NE-C and E-C and all values are reported as pg/ml (Weise and Kopin 1976).
(2) Plasma for MHPG was obtained via an indwelling catheter after 30 minutes of recumbent bed rest, stored at −70C, and was assayed by a sensitive and specific gas chromatography-mass spectrometry (GCMS) method utilizing deuterium-labeled MHPG and ^{18}O-labeled DHPG as internal standards (Jimerson 1981b).
(3) Twenty-four hour urine collections for total catecholamines and their metabolites (NE, E, MHPG, VMA, metanephrine, normetanephrine) were obtained, but values were unavailable at the time of publication.
(4) Echocardiograms and 1-mg dexamethasone-suppression tests were obtained before and after treatment and are reported elsewhere (Peterson et al., manuscripts in preparation and submitted for publication).

Statistical Analysis

Analysis of treatment response and effect of serum TCA level was performed by multiple analysis of variance (MANOVA). Comparison of categories of treatment response on the five-point scales was by χ^2 analysis. In certain instances, Pearson correlations between variables at baseline and also between baseline variables and change scores are reported. Comparison of agoraphobics and normals was by group *t*-test. Analysis of the effect of depression on treatment response and biological variables was again by MANOVA. All comparisons employed two-tailed tests of significance.

Demographics

Of the 38 patients in the beginning of the study, only two patients dropped out, one from each serum level group. These patients dropped out in the first week and were not considered in data analysis. There remained 18 patients in each of the two serum-level groups who completed 12 weeks of treatment. The final sample was 33 women and three men with a mean ± SD age of 36.7 ± 12.6 years.

The control population for the NE and E comparisons were nine normal volunteers (six women, three men) with a mean ± SD age of 37 ± 12.4 years. Since MHPG values on these controls are not currently available, a normal control population studied under similar conditions and assayed in the same laboratory was utilized. This group of 21 volunteers (20 women, five men) had a mean ± SD age of 30 ± 17 years (Siever and Uhde 1984).

RESULTS

Treatment Effects

There were highly significant treatment effects on *all* of the outcome measures. On both of the two principal outcome measures, the Symptom Severity and Avoidance Behavior Scale and the Work and Social Disability Scale, improvement was significant. Over 12 weeks of treatment, scores on the Symptom Severity and Avoidance Behavior Scale fell from 4.53 ± .65 to 2.42 ± .94 ($p < .01$), and on the Work and Social Disability Scale from 4.64 ± .64 to 2.28 ± 1.1 ($p < .01$) (Figures 1 and 2). Improvement was significant by six weeks and continued through week 12 (Figure 3).

Improvement was significant ($p < .01$) on all nine scales of the SCL-90. In particular, scores fell significantly by 12 weeks on the three most pertinent factors, the phobic anxiety (76.6 ± 5.1 to 63.0 ± 11.9), anxiety (75.6 ± 4.2 to 58.2 ± 11.7), and obsessive–compulsive factors (65.8 ± 10.4 to 52.7 ± 14.7). Additionally, improvement was significant ($p < .01$) on the Wolpe–Lange Fear Survey (132.5 ± 52.9 to 74.9 ± 40.2) at week 12.

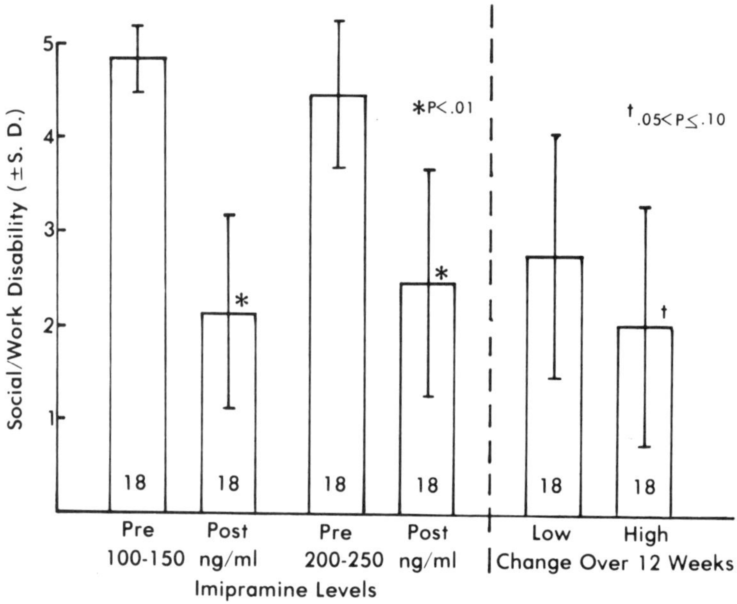

Figure 1. Marked improvement on the Symptom Severity and Avoidance Behavior Scale.

Symptomatology other than phobic anxiety and anxiety improved as well. The somatization factor on the SCL-90 fell significantly (68.2 ± 7.8 to 52.5 ± 9.6, $p < .01$). Depression improved significantly as measured on the Beck Depression Inventory (10.4 ± 6.8 to 4.5 ± 4.9, $p < .01$) and the Depression factor of the SCL-90 (69.1 ± 6.7 to 55.1 ± 12.0, $p < .01$).

Dose Effects

There were no main effects of serum TCA level (MANOVA) except that the group maintained in the 100–150 ng/ml range improved significantly more ($p < .01$) on the Symptom Severity and Avoidance Behavior Scale (Figure 4). There was also a nonsignificant trend ($p < .094$) for that group to show more improvement on the other five-point clinical scale, the Work and Social Disability Scale.

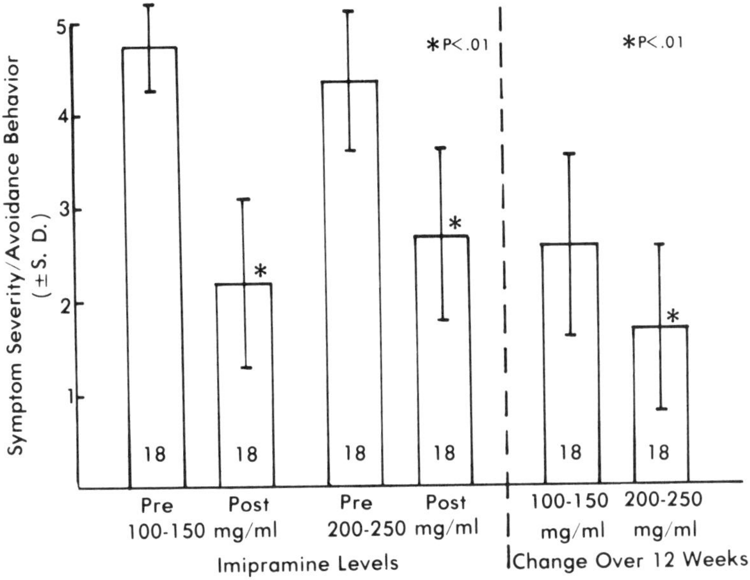

Figure 2. Marked improvement on the Work and Social Disability Scale.

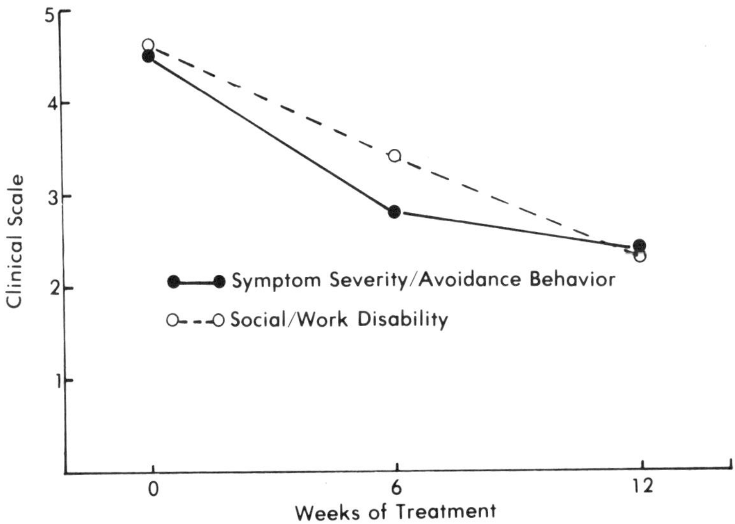

Figure 3. Time course of clinical improvement on imipramine.

Categories of Improvement

Figures 5 and 6 depict the extent of improvement on the two five-point clinical scales by categorizing improvement as marked, improved, no change, worse (see Methods). On the Work and Social Disability Scale, 77 percent of patients in both TCA-level groups had marked improvements. There were fewer marked responses on the Symptom Severity and Avoidance Behavior Scale (61 percent in the 100–150 ng/ml group, 33 percent in 200–250 ng/ml), but again the percentages of patients improved (marked plus improved) was approximately 90 percent. There was no significant difference in these categories between the two serum levels of TCA (χ^2 analysis).

Effects of Depression on Treatment Response

The 18 patients who had scores equal to or greater than 18 on their pretreatment Hamilton Rating Scale for Depression were arbitrarily classified as depressed. This is a commonly used cutoff

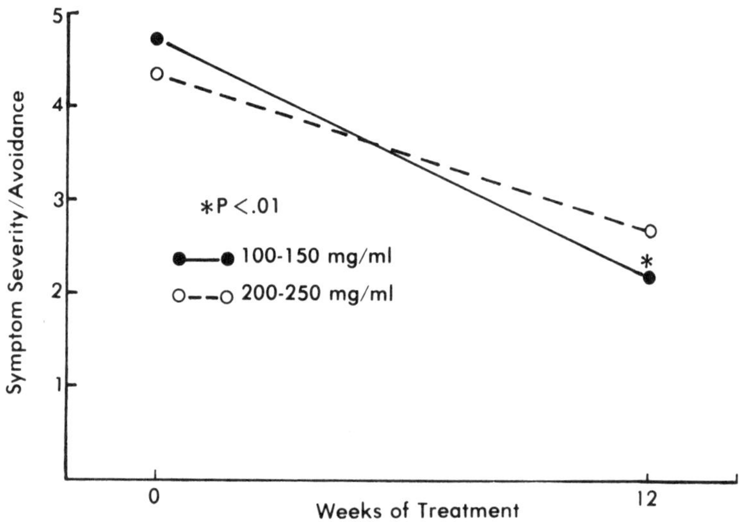

Figure 4. Lower serum level of imipramine associated with greater improvement (symptom severity/avoidance behavior).

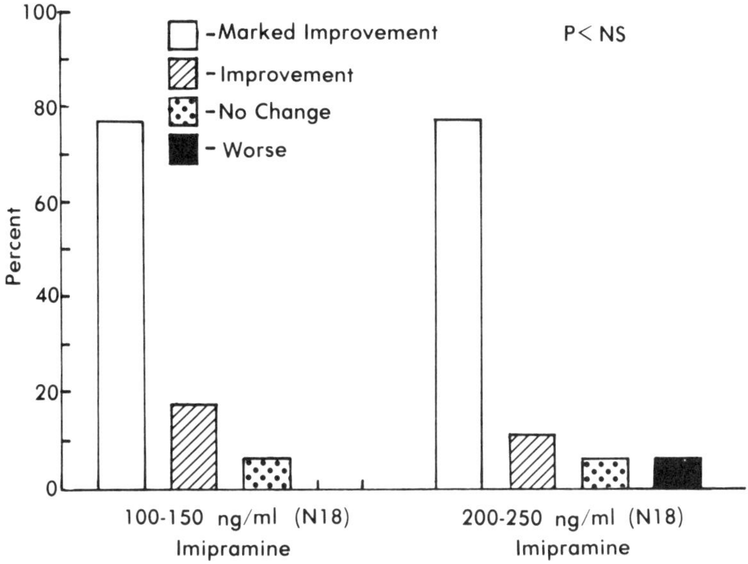

Figure 5. Categories of improvement on the Work and Social Disability Scale.

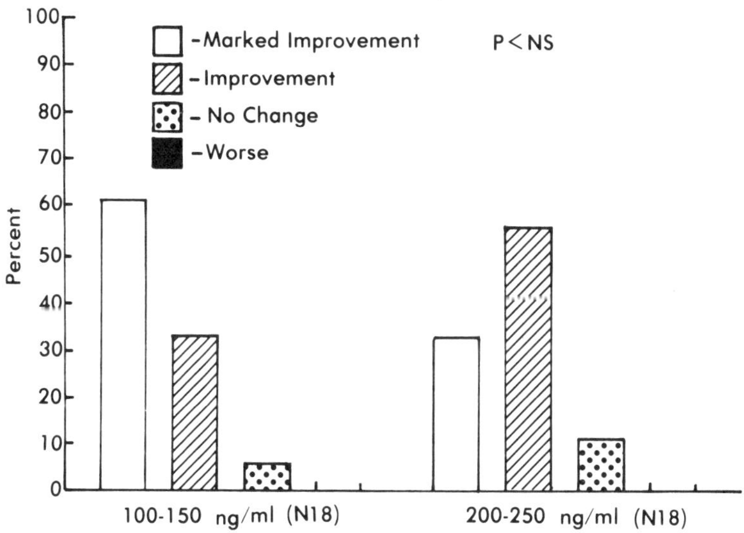

Figure 6. Categories of improvement on the Symptom Severity and Avoidance Behavior Scale.

score in studies of patients with primary depression. A high score on the Hamilton Depression Scale was not associated with either significantly increased or decreased efficacy for imipramine as measured on the two five point clinical outcome scales, the Symptom Severity and Avoidance and the Work and Social Disability Scales. Nor was it associated with significant differences on the three principal "agoraphobia" factors of the SCL-90, the phobic anxiety, anxiety, and obsessive–compulsive factors.

However, as might be expected, the depressed patients did improve more on certain other scales. They improved significantly more than the nondepressed patients on the Beck Depression Inventory ($p < .05$) and all the other ("nonphobic") factors of the SCL-90, including somatization ($p < .05$).

Higher scores on the other two measures of depression, the Beck Depression Inventory and the Depression factor from the SCL-90, were generally associated with a poor response to treatment. The BDI was negatively correlated with outcome on the Work and Social Disability Scale ($r = -0.56$, $p < .023$), the Symptom Severity and Avoidance Scale ($r = -0.64$, $p < .007$), the SCL-90 Phobic Anxiety ($r = -0.51$, $p < .046$), and the SCL-90 Total ($r = -0.52$, $p < .037$). The SCL-90 Depression factor also correlated negatively with the Work and Social Disability Scale ($r = -0.47$, $p < .065$), the Symptom Severity and Avoidance Scale ($r = -0.60$, $p < .015$), and the SCL-90 Phobic Anxiety factor ($r = -0.45$, $p < .08$).

As illustrated in Table 3, the depressed patients (HAM-D > 18) had higher pretreatment levels on the Wolpe–Lange. Their scores also fell farther with treatment, but it is difficult to rule out the possibility that this simply reflects that they had farther to fall because the percentage decreases are essentially the same in the two groups.

Table 3 Effects of Depression on Improvement on the Wolpe–Lange

Agoraphobics	(*N*)	Pretreatment	After 12 weeks	Percentage Decrease
Depression	(18)	115	88	43%
Nondepression	(18)	111	65	41%

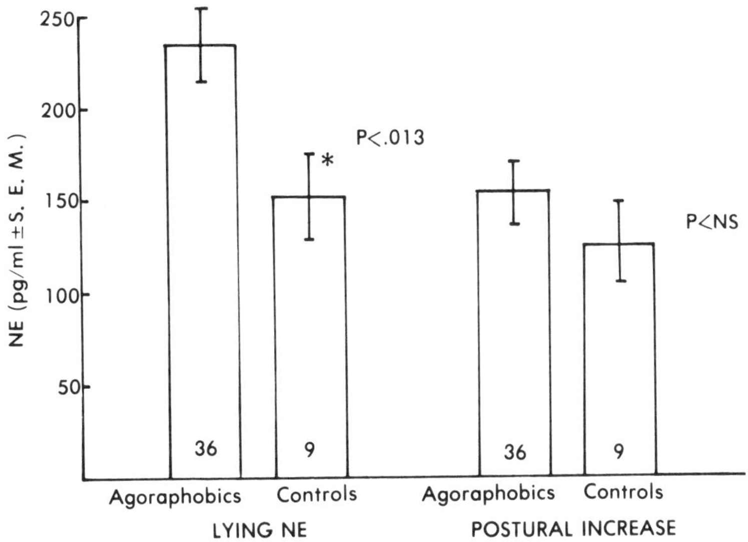

Figure 7. Increases in plasma resting (supine) levels of norepinephrine (NE) in agoraphobics.

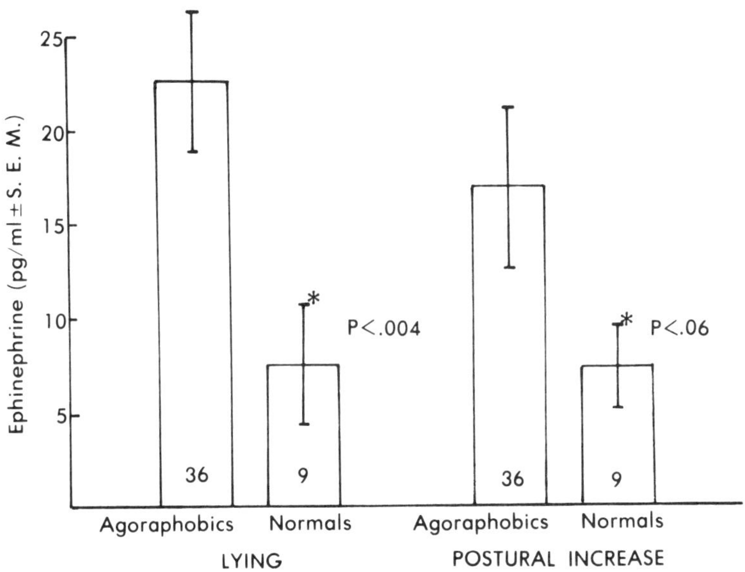

Figure 8. Increases in plasma levels of epinephrine in agoraphobics.

Biological Measures

The mean plasma NE (supine, NE-S) in the 36 patients was significantly higher ($p < .013$) than in the nine normals, although the increases on standing (NE-C) in the two groups were not significantly different (Figure 7). Plasma E levels, both supine (E-S) and the increase on standing (E-C), were significantly higher in the agoraphobic patients (Figure 8).

Mean plasma MHPG (supine) levels ($3.82 \pm .81$, $N = 12$) were also significantly ($p < .003$) higher than control values ($3.0 \pm .50$, $N = 21$).

To explore whether baseline NE and E measures were related to symptomatology, Pearson correlations were attempted between the principal psychometric measures and the four NE and E measures and plasma MHPG. There were no statistically significant relationships between these biological measures and psychometric measures of relevant symptomatology. The Work and Social Disability Scale ($r = .45$, $p < .072$) and the SCL-90 Phobic Anxiety factor did correlate with plasma MHPG at close to significance levels ($r = .49$, $p < .053$).

The BDI score correlated with the NE-C ($r = .30$, $p < .037$) as did the HAM-D with MHPG ($r = .52$, $p < .041$). Of interest in relationship to other work connecting increased NE measures to aggression (Brown et al. 1979), the SCL-90 Hostility factor correlated with MHPG ($r = .64$, $p < .012$) and E-C ($r = .33$, $p < .026$).

Catecholamines

We were not able to predict from the four baseline catecholamine measures (NE-S, NE-C, E-S, E-C) the clinical response as measured by the two five-point scales. Although scattered relationships with other measures were observed, it is doubtful that they are meaningful predictors of improvement.

Psychometric Measures

There were also very few pretreatment psychometric measures associated with a positive response at week 12. However, in general, thc more severely ill patients tended to experience more improvement. There were positive correlations between the pretreatment scores on the Work and Social Disability Scale and reductions in scores at 12 weeks on the SCL-90 Total score ($p <$ .072), Symptom Severity Scale ($p < .039$), and the Work and Social Disability Scale itself ($p < .069$).

However, the level of obsessive–compulsive features as measured by the SCL-90 Obsessive–Compulsive factor was significantly associated with negative outcome. It was negatively correlated with the 12-week Work and Social Disability ($p < .005$) and Symptom Severity and Avoidance Behavior Scales ($p < .019$), and the SCL-90 Phobic Anxiety factor ($p < .004$) and SCL-90 Total ($p < .065$).

As previously mentioned, pretreatment measures of depression were either not predictive of response or generally had negative associations with treatment response.

Biological Changes with Treatment

Norepinephrine, both NE-S and NE-C, increased significantly with imipramine treatment (Figure 9). However, E-S significantly decreased ($p < .01$) with treatment (22.7 $\pm$ 22.9 to 14.3 $\pm$ 13.3); E-C did not change significantly. Plasma MHPG significantly

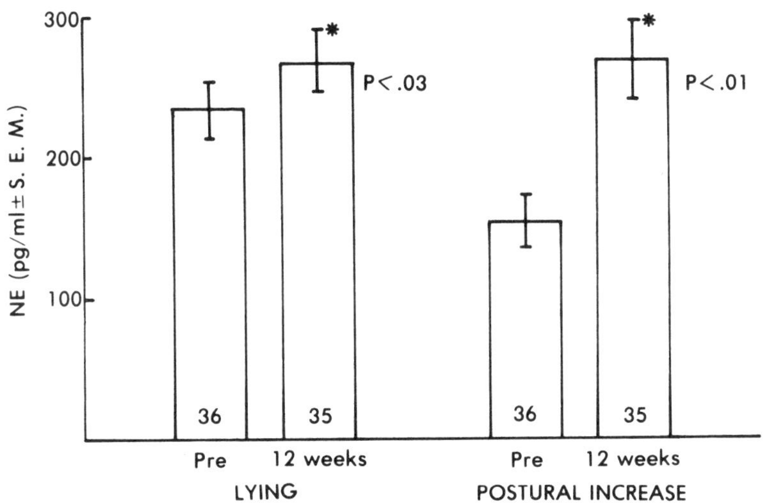

Figure 9. Increases in plasma norepinephrine (NE) with imipramine treatment.

decreased from 3.82 ± 0.81 to 2.64 ± .79 ($p < .001$, N = 12) at week 12. MHPG at that point (12 weeks) was not significantly different from the control group.

There were no significant main effects of tricyclic serum level on any of these biological variables.

Biological Changes in Depressed Agoraphobics

There were no differences in the observed changes in NE and E with treatment in the agoraphobics who had baseline HAM-D scores above and below the cutoff of 18.

COMMENT

Clinical Response

The clinical response of patients in this study was excellent. Most of the patients had a marked response to three months of treatment with imipramine and weekly group therapy that was supportive and promoted in vivo exposure ("homework"). This response was observed not only on psychometric scales (e.g., SCL-90, Wolpe–Lange), but also on the two five-point clinical scales. These scales are global clinical scales with changes based on actual behavioral changes. These scales have been useful in previous research to document the clinical meaningfulness of improvement in agoraphobic patients. Most of the patients felt that their agoraphobic symptomatology was essentially resolved. This was documented by 77 percent of the patients having marked responses on the Work and Social Disability Scale. We feel that this scale is the most significant indicator of improvement since it is based on actual decreases in disability in the work and social behavior of these patients.

The improvement on the Work and Social Disability Scale is important for another reason. Cognitively, most, if not all, of these patients are characterized by a "*what if*" attitude, which leads not only to their phobic avoidance behavior, but also to a particular attitude toward their improvement. They are extremely skeptical that any decrease in panic attacks or anxiety is real, feeling "yes, but the symptoms could come back at any time. I've gotten better for a short time before." It is perhaps this attitude that is mirrored in our study by the patients' reporting less improvement in anxiety symptomatology and avoidance behavior than they do in improvement in actual behavioral disability in the work and social areas. That is, they are actually doing better functionally than they feel they are doing symptomatically. This clinical characteristic further argues that the highly significant improvement on the Work and Social Disability Scale is the best indicator of the excellent clinical response to this treatment at three months. The "what if" nature of the anxiety and avoidance behavior leads it to actually resolve more slowly than impairment in behavioral areas. We will observe in longer term follow-up if improvement on the Symptom Severity and Avoidance Behavior Scales "catches up" with the improvement on the Work and Social Disability Scale.

Improvement on most indices was significant by week six of the trial, but there were significant clinical gains from weeks six to 12 as well.

The response of agoraphobic patients was not restricted to improvement in phobic symptomatology. As a group, they also had significant decreases on the depression scales (BDI and SCL-90, Depression factor), all of the SCL-90 scales, and the SCL-90 Total Psychopathology Score. Therefore, they had significant improvement on the obsessive–compulsive scales, somatization, and the other psychopathology scales (paranoid ideation, psychoticism, interpersonal sensitivity, etc.). This general improvement is consistent with previous observations of imipramine's "patholytic" effect in agoraphobics. As individuals, the change, at times, can appear to be from an extremely severe psychopathological state to a "normal" state.

Anxiety was significantly reduced by treatment, as reflected by the significant decrease on the SCL-90 Anxiety factor. This decrease presumably reflects a decrease in the prominent *anticipatory anxiety* in agoraphobics. This term was coined by Donald Klein for the pervasive anxiety of agoraphobics, anxiety that is differentiated from the anxiety of the panic attack or the anxiety experienced in the phobic situation. It is thought to be the anxiety (or worry) in anticipation that another panic attack will occur. Some clinicians have felt this type of anxiety is not reduced by TCAs and requires benzodiazepine co-treatment. Our results replicate an earlier finding that treatment with imipramine or phenelzine alone results in significant reduction of the SCL-90 Anxiety factor (Sheehan et al. 1980).

This is the tenth reported controlled study of imipramine in this condition, and most have demonstrated a clear, significant, and positive response. Although still controversial, in our opinion it is established that imipramine is highly effective in treating agoraphobia, presumably by blocking panic attacks. The most important questions that remain concern its relationship to behavior therapy. Direct comparisons between behavior therapy and imipramine alone, and imipramine combined with behavior therapy need to be studied at this point. The direct comparisons to date leave room for interpretation and controversy. The studies of Zitrin, Klein, and colleagues (Zitrin et al. 1983) and Sheehan, Ballenger, and Jacobsen (Sheehan et al. 1980) provide evidence that imipramine leads to more significant improvement than behavior therapy techniques alone. However, Marks's study (Marks et al. 1983) reports the opposite conclusion.

The methodology of this study did not allow investigation of imipramine versus behavior therapy, or of therapy with imipramine alone, or either compared to placebo. We treated all patients with imipramine and group therapy based on behavioral principles. We are currently studying the effectiveness of medications and only brief exposure instructions, and direct comparisons of medications and behavioral treatments utilized alone and in combination are being planned in other centers.

A related question is the apparent differential relapse rate with the two types of therapy. The reported relapse rate for antidepressant treatment of agoraphobia varies from 17 to 80 percent (Zitrin et al. 1978; Sheehan 1983) with less than 15 percent relapsing after behavior therapy (Marks 1981). Even these studies are unclear because of failure to define relapse and the different lengths of follow-up. Also, relapse may appear to be more significant in the medication trials because of a greater "quantity" of improvement, with greater opportunity to relapse in the medication groups. These questions require further study as part of the evaluation of the differential value of the two types of treatment.

We failed to observe any clear difference in treatment response between the group maintained at the 100–150 ng/ml serum level and the one at 200–250 ng/ml. However, there was a tendency for the lower level group to respond better clinically on the two important clinical scales. Because of this trend, we hesitate at this point to definitively conclude that there is no relationship between serum level and clinical response. Further studies are indicated with larger sample sizes and with the use of different methodologies to evaluate the relationship of serum TCA to clinical response. It is certainly possible that further studies may definitively demonstrate superior response in the lower serum range; however, it would be our clinical recommendation even at this point, based on our data, that if plasma levels are utilized, patients should be maintained in the 100–150 ng/ml range. Patients in that range had excellent clinical responses, and there was no evidence that the average patient should be maintained at higher levels. However, if any single patient fails to respond at that level after an adequate length of time (four to six weeks), it would be clinically reasonable to increase doses. Our results and recommendations are, of course, somewhat inconsistent with the data from Mavissakalian and colleagues who found better responses to higher serum levels of imipramine (Mavissakalian et al. 1984). Their sample was small ($N = 15$) and they utilized a different methodology, but there is no ready explanation for the differences observed.

There were also no differences in the changes in NE and E with treatment between the groups. It appears that the biological changes are accomplished at the lower level and higher levels make no additional difference.

Some have argued that the efficacy of antidepressants in this condition is based solely on the improvement of phobic patients who are also depressed (or dysphoric) (Marks 1981). However, our patients who had Hamilton depression scores above the cutoff score of 18 at baseline had the same response as the "nondepressed" (HAM-D less than 18) agoraphobics. Also the patients who had indications of higher depression on the self-report measures (BDI and SCL-90) actually experienced less improvement in areas of phobic symptomatology and disability. These results are similar to previous research demonstrating either no difference in improvement for the depressed agoraphobics (Sheehan et al. 1980; Zitrin et al. 1983) or actually less favorable responses (Zitrin 1980).

It is interesting to note that symptomatology other than specific phobic symptomatology improved more in the depressed agoraphobics. As expected, the depressed patients experienced greater decreases in depression, but they also improved more on all the nonphobic scales of the SCL-90. We might expect the depressed agoraphobics to have greater somatic symptomatology and greater nondepressive, nonphobic psychopathology as well. These other symptoms do improve significantly in these depressed patients, perhaps more than in the nondepressed patients, because they were initially greater and therefore have farther to fall.

We had hoped to identify biological or psychometric indices that would predict improvement with this treatment. We were able to identify only one significant positive relationship. Higher pretreatment scores on the Work and Social Disability Scale did predict improvement as measured by the SCL-90 Total, Symptom Severity Scale, and the Work and Social Disability Scale itself. This may reflect a general trend for the more severely ill agoraphobic patients to respond more significantly. We cannot determine whether the more severely ill truly do respond better to this treatment, or whether this relationship simply reflects a mathematical relationship, that they have "greater room" for their scores to fall.

The agoraphobic patients with high obsessive–compulsive features, as measured by the SCL-90 Obsessive–Compulsive factor, did significantly *less* well, and this probably reflects a true negative predictor. The negative relationship of the SLC-90 Obsessive–Compulsive factor to the main outcome measures (two five-point clinical scales and the SCL-90 phobic anxiety and SCL Total) was highly significant. Although obsessive–compulsive scores were significantly decreased in these patients, as they were in the study by Sheehan and colleagues (Sheehan et al. 1980), their presence seems to limit improvement in the phobia measures.

The biological measures were also examined for predictive value (see Chapter 3, as well) with the hope that we might be able to identify an easily obtainable biological measure associated with improvement. With one possible exception, we did not observe any relationships between the pretreatment biological measures and clinical outcome that appeared to be clinically meaningful. The increase in E on standing (at pretreatment) was correlated with positive improvement on the SCL-90 Total and Somatization factors, and at a trend level on the Wolpe–Lange. However, even this relationship was not observed on the more central phobia outcome measures.

Biological Measures

The mean plasma NE level of the agoraphobic patients was significantly higher in the agoraphobic patients than in normals, although the increase with postural change was not. Most of the plasma NE is thought to be derived from the peripheral sympathetic system, and the postural increase to reflect the reactivity and integrity of that system. We might speculate that these increased levels of basal NE reflect increased central "drive" of this system. This increase in NE is certainly consistent with our original ideas that agoraphobics have an increase in the "set point" for the NE system. Since the agoraphobics who panic with lactate have higher pre-lactate NE levels, perhaps NE levels are important in establishing the "set point" for panic as well.

The increases we observed in supine NE, although significant, are relatively modest in size. This magnitude of increase could well be obscured by the large increment seen on postural change and may explain why we did not observe significant differences in NE between agoraphobics and controls with postural change. That the mean plasma MHPG values were also significantly higher than in normal controls is again consistent with increased activity of the central NE system, including the LC, and of the peripheral sympathetic nervous system.

As mentioned previously, the relationship between central and peripheral NE systems remains unclear, although evidence does suggest that they work in concert with increased central (primarily LC) NE release resulting in increased peripheral release. Evidence suggests that LC firing increases MHPG of many brain regions, and since free MHPG readily diffuses out into cerebrospinal fluid (CSF) and blood, there are high correlations between brain, CSF, and plasma MHPG in primates (Elsworth et al. 1982). We have also previously reported a significant correlation in normal humans between plasma MHPG and CSF MHPG (Jimerson et al. 1981a). These brain–plasma correlations presumably result from direct or indirect central control of peripheral MHPG production, coupled with an increase in plasma MHPG secondary to increased outflow of MHPG from brain areas with projections from the LC. The stability and long half-life of MHPG also contribute to its usefulness as a marker of central NE activity.

Several lines of evidence are available that link CNS NE activity and phobic anxiety with plasma measures of NE metabolism, most notably MHPG. As mentioned, plasma MHPG increases during panic attacks and this increase is blocked by both clonidine and imipramine (Ko et al. 1983). Both these agents decrease LC firing (Svensson et al. 1975; Nyback et al. 1975) and clonidine has been shown to decrease CNS NE turnover (Anden et al. 1970) and brain MHPG (Maas et al. 1977). Both clonidine and imipramine decrease plasma MHPG (Ko 1983; Leckman et al. 1980) and are effective against panic anxiety. Piperoxane and yohimbine, which have the opposite effects of increasing LC firing and anxiety (Holmberg and Gershon 1961), also increase both CNS and plasma MHPG (Elsworth et al. 1982; Maas et al. 1977).

We feel the evidence we have reviewed suggests important linkages between CNS NE activity and plasma noradrenergic measures and supports our use of plasma measures as reflections of CNS NE activity. When complete plasma, urinary, and cerebrospinal fluid catecholamine data are available in these patients, this relationship can and will be further explored. Since metabolism of NE to metabolites other than MHPG varies between individuals and is influenced by medications, analysis of the pattern and amounts of all the metabolites (total body catecholamines) in the urine of these patients will be particularly useful. However, the data we report are consistent with our original hypothesis that agoraphobic patients have increased catecholamine release, based perhaps on increased activation of the CNS NE system. However, these catecholamine measures were not significantly related to pretreatment symptomatology, except at a trend level for plasma MHPG and the Phobic Anxiety factor of the SCL-90.

Although the differences between patients and normals in these NE and E measures were statistically significant, they were not large differences. Also the limits of the accuracy of the assays for NE and E are approximately 30 pg/ml for both NE and E and make our data preliminary and in need of replication.

Both NE measures increased with treatment. These increases have been previously observed with imipramine treatment of depressed patients (Ross et al. 1983; Veith et al. 1983). In these studies, E increased in a parallel fashion. In our agoraphobic patients, E decreased with treatment, with the decrease in the supine (resting) E levels being significant. Since most of the plasma E is adrenal in origin and reflects the state of arousal and anxiety, the decreases in E probably are secondary to the improvement in anxiety symptomatology these patients experienced. Plasma MHPG levels also fell significantly with treatment. This is similar to the decreased MHPG observed with chronic desmethylimipramine in depressed patients (Charney et al. 1981b). These decreases in MHPG are consistent with decreases in central and peripheral turnover of NE. Although speculative, such decreases could also be expected if CNS NE (and LC) activity were reduced by treatment. These data are consistent with the evidence that TCAs decrease firing of the LC and that this decrease in LC activity is at least part of the mechanism of their antipanic, antiphobic activity.

To understand these changes with treatment it is important to review, at least briefly, the complex evidence of the actions of the tricyclics like imipramine. Certainly their earliest and best known action is to block the reuptake of NE released into the synapse (Carlsson et al. 1969; Paton 1976). Since this is the primary mechanism for the deactivation of released NE, the net effect has been felt to be an increase in NE available in the synapse. More recent work has demonstrated that tricyclics result in a desensitization of presynaptic alpha-2 receptors (Crews and Smith 1978), perhaps via a decrease in the number of receptors (Smith et al. 1981).

Turnover studies have been contradictory, with some demonstrating decreases and others increases in turnover (Maas et al. 1972; Schildkraut 1975; Charney et al. 1981a).

Recent studies demonstrating decreases in whole-body turnover (Linnoila et al. 1982) are most pertinent to our results and argue strongly that the net (whole-body) effect of TCAs is to decrease NE activity. This is certainly consistent with the evidence showing that TCAs decrease LC firing and with the current understanding of the LC's central role in the NE system. Multiple studies document decreases in plasma, urinary, and CSF MHPG with TCA treatment (Potter et al. 1981; Schildkraut et al. 1964; Maas et al. 1972). This understanding of TCA action is then consistent with the decreased MHPG we observed and with the action of TCAs we hypothesized (i.e., to decrease abnormally high NE activity in agoraphobics).

The action of TCAs to block reuptake is felt to increase "spillage" of NE from synapses in the peripheral sympathetic nervous system (Linnoila et al. 1982). Also, one dose of desipramine has been reported to decrease clearance by 20 percent (Esler et al. 1981). Although certainly open to argument (Veith et al. 1983; Ross et al. 1983), we feel the preponderance of evidence suggests that the overall activity of TCAs on NE is to decrease total-body turnover and that the increases in plasma NE we and others have observed (Veith et al. 1983; Ross et al. 1983) may well be secondary to spillage and decreased clearance. If this is the preponderant effect of TCAs on plasma NE, it would certainly reduce its value as a marker of total NE activity *after* TCA treatment (not before). However, the significant increases in the postural changes for NE argues that the imipramine effect on NE may not simply be on clearance.

The evidence of increased norepinephrine activity in agoraphobics and the decreases in some of these measures into the normal range on imipramine are consistent with the hypothesized involvement of norepinephrine and the LC in agoraphobia. It is important to emphasize, however, that it is unlikely that this hypothesis is a complete explanation of this complex syndrome. For instance, the beta-adrenergic blocker propranolol does not block panic attacks elicited by lactate (Gorman et al. 1983). Other models have been proposed (see Chapters 4 and 5), and this area is under active investigation. A more definitive delineation of the role of norepinephrine, the LC, and other brain mechanisms in panic anxiety awaits further scientific investigation.

References

Anden NE, Corrodi H, Fuxe K, et al: Evidence for a central noradrenaline receptor stimulation by clonidine. Life Sci 9:513–523, 1970

Ballenger JC, Post RM, Jimerson DC, et al: Neurobiological correlates of depression and anxiety in normal individuals in Neurobiology of Mood Disorders. Edited by Post RM, Ballenger JC. Baltimore, Williams & Wilkins, 1984, pp 481–501

Beck AT, Beamesderfer A: Assessment of depression: the depression inventory, in Psychological Measurements in Psychopharmacology, vol 7. Edited by Pichot P. Basel, Karger, 1974, pp 151–169

Bertrand M: The Behavioral Repertoire of the Stump Tail Macaque. New York, S. Karger, 1969, pp 95, 98–99, 175

Bliss E, Ailion J: Response of neurogenic amines to aggregation and strangers. J Pharmacol Exp Ther 168:258, 1969

Bloom G, Euler US, Frankenhauser M: Catecholamine excretion and personality traits in paratroop trainees. Acta Physiol Scand 58:77–89, 1963

Boudoulas H, Reynolds JC, Mazzaferri E, et al: Metabolic studies in mitral valve prolapse syndrome: a neuroendocrine–cardiovascular process. Circulation 61:1200–1205, 1980

Brown GL, Goodwin FK, Ballenger JC, et al: Aggression in humans correlates with cerebrospinal fluid amine metabolites. Psychiatry Res 1:131–139, 1979

Carlsson A, Corrodi H, Hokfelt K, et al: Effects of some antidepressant drugs on the depletion of intraneuronal brain catecholamine stress caused by 4-α-dimethylmeta-tyramine. Eur J Pharmacol 5:357–366, 1969

Charney DS, Menkes DB, Heninger GR: Receptor sensitivity and the mechanism of action of antidepressant treatment: implications for the etiology and therapy of depression. Arch Gen Psychiatry 38:1160–1180, 1981a

Charney DS, Heninger GR, Sternberg DF, et al: Plasma MHPG in depression: effects of acute and chronic desipramine treatment. Psychiatry Res 5:217–229, 1981b

Crawley JN, Hattox SE, Maas JW, et al: 3-Methoxy-4-hydroxyphenethylene glycol increase in plasma after stimulation of the locus coeruleus. Brain Res 141:380–384, 1978

Crews FT, Smith CB: Presynaptic alpha-receptor subsensitivity after long-term antidepressant treatment. Science 202:332–334, 1978

DeMet EM, Halaris AE: Origin and distribution of 3-methoxy-4-hydroxyphenyl glycol in body fluids. Biochem Pharmacol 28:3042–3050, 1979

Derogatis LR, Lipman RS, Rickels K, et al: The Hopkins Symptom Checklist (HSCL): a self-report symptom inventory. Behav Science 19(1):1–15, 1974

DuPont RL: Profile of a phobia treatment program: two year follow-up, in Phobia: A Comprehensive Summary of Modern Treatments. Edited by DuPont RL. New York, Brunner/Mazel, 1982

Easton JD, Sherman DG: Somatic anxiety attacks and propranolol. Arch Neurol 33:689–691, 1976

Edmonson HD, Roscoe B, Vickers MD: Biochemical evidence of anxiety in dental patients. Br Med J 4:7–9, 1972

Elsworth JD, Redmond DE Jr., Roth RH: Plasma and cerebrospinal fluid 3-methoxy-4-hydroxy phenylethylene glycol (MHPG) as indices of brain norepinephrine metabolism in primates. Brain Res 235:115–124, 1982

Esler M, Jackman G, Leonard P, et al: Effect of norepinephrine uptake blockers on norepinephrine kinetics. Clin Pharmacol Ther 29:12–20, 1981

Gold MS, Redmond DW Jr: Pharmacological activation and inhibition of noradrenergic activity alter specific behaviors in nonhuman primates. Neuroscience 3:250, 1977

Gordon R, Spector S, Sjoerdsma A, et al: Increased synthesis of norepinephrine and epinephrine in the intact rat during exercise and exposure to cold. J Pharmacol Exp Ther 153:440–447, 1966

Gorman JM, Levy GF, Liebowitz MR, et al: Effect of acute adrenergic blockage on lactate-induced panic. Arch Gen Psychiatry 40: 1079–1082, 1983

Hamilton M: A rating scale for depression. J Neurol Neurosurg Psychiatry 23:56–62, 1960

Holmberg G, Gershon S: Autonomic and psychic effects of yohimbine hydrochloride. Psychopharmacologia 2:93–106, 1961

Huang HY, Maas JW, Redmond DE Jr: Evidence for noradrenergic specificity of behavioral effects of electrical stimulation of the nucleus locus coeruleus. Neuroscience Abstracts 3:251, 1977

Huang HY, Redmond DE, Snyder DR, et al: Loss of fear following bilateral lesions of the locus coeruleus in the monkey. Neuroscience Abstracts 2:573, 1976

Jimerson DC, Ballenger JC, Lake CR, et al: Plasma and CSF MHPG in normals. Psychopharmacol Bull 17:86–87, 1981a

Jimerson DC, Markey SP, Oliver JO, et al: Simulanteous measurement of plasma 4-hydroxy-3-methoxy phenylethylene glycol and 3, 4-dihydroxphenylethylene glycol by gas chromatography-mass spectrometry. Biomed Mass Spectrom 8:256–259, 1981b

Jobson K, Linnoila M, Gillam J, et al: Successful treatment of severe anxiety attacks with tricyclic antidepressants: a potential mechanism of action. Am J Psychiatry 135:863–865, 1978

Jones MT, Bridges PK, Leak D: Relationship between the cardiovascular and sympathetic response to the psychological stress of an examination. Clin Sci 35:73–79, 1968

Kantor JS, Zitrin CM, Zeldis SM: Mitral valve prolapse syndrome in agoraphobic patients. Am J Psychiatry, 137:467–469, 1980

Kelly D, Guirguis W, Frommer E, et al: Treatment of phobic states with antidepressants. Br J Psychiatry 116:387–398, 1970

Klein DF: Delineation of two drug-responsive anxiety syndromes. Psychopharmacologia 53:397–408, 1964

Klein DF: Importance of psychiatric diagnosis in prediction of clinical drug effects. Arch Gen Psychiatry 16:118–125, 1967

Klein DF, Fink M: Psychiatric reaction patterns to imipramine. Am J Psychiatry 119:432–438, 1962

Ko GN, Elsworth JD, Roth RH, et al: Panic-induced elevation of plasma MHPG levels in phobic-anxious patients. Arch Gen Psychiatry 40: 425–430, 1983

Leckman JF, Maas JW, Redmond DE, et al: Effect of oral clonidine on plasma 3-methoxy-4-hydroxy phenylethylene glycol in man. A preliminary report. Life Sci 26:2179–2185, 1980

Linnoila M, Karoum F, Calil H, et al: Alteration of norepinephrine metabolism with desipramine and zimelidine in depressed patients. Arch Gen Psychiatry 39:1025–1028, 1982

Maas JW, Fawcett JA, Dekirmenjian H: Catecholamine metabolism, depressive illness, and drug response. Arch Gen Psychiatry 26: 252–262, 1972

Maas JW, Hattox SE, Landis DH, et al: A direct method for studying 3-methoxy-4-hydroxy-phenylethylene glycol (MHPG) production by brain in awake animals. Eur J Pharmacol 46:221–228, 1977

Marks IM: Fears and Phobias. New York, Academic Press, 1969

Marks IM: 'Are there anticompulsive or antipanic drugs?' A review of the evidence. Presented at the American College of Neuropharmacology (ACNP), San Diego, December 18, 1981

Marks IM, Birley MLT, Gelder MG: Modified leucotomy in severe agoraphobia: a controlled serial inquiry. Br J Psychiatry 112:757–769, 1966

Marks IM, Gray S, Cohen D, et al: Imipramine and brief therapist-aided exposure in agoraphobics having self-exposure homework. Arch Gen Psychiatry 40:153–162, 1983

Mathew RJ, Ho BT, Kralik P, et al: Catechol-o-methyltransferase and catecholamines in anxiety and relaxation. Psychiatry Res 3:85–91, 1980

Mavissakalian M, Perel J, Michelson L: The relationship of plasma imipramine and N-desmethylimipramine to improvement in agoraphobia. J Clin Psychopharmacol 4(1):36–40, 1984

McNair DM, Kahn RJ: Imipramine compared with a benzodiazepine for agoraphobia, in Anxiety: New Research and Changing Concepts. Edited by Klein DF, Rabkin JG. New York, Raven Press, 1981, pp 69–80

Nies A, Robinson D: New studies confirm MAO inhibitors' efficacy in treating severe anxiety. JAMA 245:1799–1801, 1981

Nyback H, Walters JR, Aghajanian GK, et al: Tricyclic antidepressants: effects on the firing rate of brain noradrenergic neurons. Eur J Pharmacol 32:302–312, 1975

Paton DM: The mechanism of neuronal and extraneuronal transport of catecholamines. New York, Raven Press, 1976, pp 95–153, 325–354

Pitts FN, Allen RE: Biochemical induction of anxiety, in Phenomenology and Treatment of Anxiety. Edited by Fann WE, Karacan I, Pokorny AD, et al. New York, Spectrum Publications, 1979, pp 125–150

Post RM, Lake CR, Jimerson DC, et al: Cerebrospinal fluid norepinephrine in affective illness. Am J Psychiatry 135:907–917, 1978

Potter WZ, Calil HM, Extein I, et al: Specific norepinephrine and serotonin inhibitors in man: a crossover study with pharmacokinetic, biochemical, neuroendocrine, and behavioral parameters. Acta Psychiatr Scand [Suppl 290] 63:152–164, 1981

Redmond DE Jr: Alterations in the function of the nucleus locus coeruleus: a possible model for studies of anxiety, in Animal Models in Psychiatry and Neurology. Edited by Hanin I, Usdin, E. Oxford, Pergamon Press, 1977

Redmond DE Jr: New and old evidence for the involvement of a brain norepinephrine system in anxiety, in Phenomenology and Treatment of Anxiety. Edited by Fann WE, Karacan I. New York, Spectrum Publications, 1979, pp 153–203

Redmond DE, Huang HY: II. New evidence for a locus coeruleus–norepinephrine connection with anxiety. Life Sci 25:2149–2162, 1979

Redmond DE Jr., Huang HY, Snyder DR, et al: Behavioral effects of stimulation of locus coeruleus in the stumptail monkey (*Macaca arctoides*). Brain Res 116:502–510, 1976a

Redmond D, Huang Y, Snyder D, et al: Behavioral changes following lesions of the locus coeruleus in macaca arctoids. Neuroscience 1:472, 1976b

Ross RJ, Zavadil AP, Calil HM, et al: Effects of desmethylimipramine on plasma norepinephrine, pulse, and blood pressure. Clin Pharmacol Ther 33:429–437, 1983

Rubin RT, Miller RG, Clar BR, et al: The stress of aircraft carrier landings II. 3-methoxy-4-hydroxy-phenylethylene glycol excretion in naval aviators. Psychosom Med 32:589–597, 1970

Schildkraut JJ: Norepinephrine metabolism after short- and long-term administration of tricyclic antidepressant drugs and electronconvulsive shock, in Neurobiological Mechanisms of Adaptation. Edited by Mandell A. New York, Raven Press, 1975

Schildkraut J, Klerman G, Hammon R, et al: Excretion of 3-methoxy-hydroxymandelic acid (VMA) in depressed patients treated with antidepressant drugs. J Psychiatr Res 2:257–266, 1964

Sheehan D: Psychopharmacological treatment of agoraphobia. Unpublished presentation at the "Biology of Panic-Related Disorders" Meeting, Boston, November 4-6, 1983

Sheehan DV, Ballenger J, Jacobsen G: Treatment of endogenous anxiety with phobic, hysterical, and hypochondriacal symptoms. Arch Gen Psychiatry 37:51–59, 1980

Siever LJ, Uhde TW: New studies and perspectives on the noradrenergic receptor system in depression: effects of the alpha-2-adrenergic agonist clonidine. Biol Psychiatry (in press)

Smith CB, Garcia-Sevilla JA, Hollingsworth PJ: α_2-Adrenoreceptors in rat brain are decreased after long-term tricyclic antidepressant drug treatment. Brain Res 210:413–418, 1981

Smith DE, Wesson DR (eds.) The benzodiazepines: two decades of research and clinical experience. Psychoactive Drugs, January–June, 1983

Stern RS, Marks IM, Wright J, et al: Clomipramine: plasma levels, side effects and outcome in obsessive compulsive neurosis. Postgraduate Med J 56(1):134–139, 1980

Stone EA: Stress and catecholamines, in Catecholamines and Behavior. Edited by Friedman AJ. New York, Plenum, 1975

Svensson TH: Effect of chronic treatment with tricyclic antidepressant drugs on identified brain noradrenergic and serotonergic neurons. Acta Psychiatr Scand [Suppl] 280:121–131, 1980

Svensson TH, Bunney BS, Aghajanian GK: Inhibition of both noradrenergic and serotonergic neurons in brain by the alpha adrenergic agonist clonidine. Brain Res 92:291–306, 1975

Sweeney DR, Maas JW, Heninger GR: State anxiety, physical anxiety, and urinary 3-methoxy-4-hydroxy-phenylethylene glycol excretion. Arch Gen Psychiatry 34:1418–1423, 1978

Thierry AM, Javoy F, Glowinski J, et al: Effects of stress on the metabolism of norepinephrine, dopamine and serotonin in the central nervous system of the rat. I. Modifications of norepinephrine turnover. J Pharmacol Exp Ther 163:163–171, 1968

Veith RC, Raskind MA, Barnes RF, et al: Tricyclic antidepressants and supine, standing, and exercise plasma norepinephrine levels. Clin Pharmacol Ther 33:763–769, 1983

Weise VK, Kopin IJ: assay of catecholamines in human plasma: studies of a single isotope radioenzymatic procedure. Life Sci 19:1673–1686, 1976

Wolpe J, Lange PJ: A fear survey schedule for use in behavior therapy. Behav Res Ther 2:27–30, 1964

Wyatt RJ, Portnoy B, Kupfer DJ, et al: Resting plasma catecholamine concentrations in patients with depression and anxiety. Arch Gen Psychiatry 24:65–70, 1971

Zitrin CM, Klein DF, Woerner MG: Behavior therapy, supportive psychotherapy, imipramine and phobias. Arch Gen Psychiatry 35:307–316, 1978

Zitrin CM, Klein DF, Woerner MG: Treatment of agoraphobia with group exposure in vivo and imipramine. Arch Gen Psychiatry 37:63–72, 1980

Zitrin CM, Klein DF, Woerner MG, et al: Treatment of phobias I. Comparison of imipramine hydrochloride and placebo. Arch Gen Psychiatry 40:125–138, 1983

3

The Psychophysiology of Agoraphobia

William R. Hobbs, M.D.
Clinton Wilkinson, Ph.D.
Gregory A. Peterson, M.D.
Michele Laraia, R.N., M.S.N.
Daniel Cox, Ph.D.
James C. Ballenger, M.D.

3

The Psychophysiology of Agoraphobia

Agoraphobia is a chronic and progressive disorder with both biological and psychological aspects. Previously under-reported, it has more recently been found to be a significant cause of morbidity in the population at large. It may be particularly over-represented in cardiology practices among patients with atypical chest pain. Although temporary remissions are seen in a minority of patients (Marks and Herst 1970), improvement without treatment is less common than a gradual worsening of the overall symptom picture.

There are anecdotal reports of patients with agoraphobia who are more likely to panic or to become anxious under fluorescent than under incandescent lighting conditions. In this study we addressed the autonomic and musculoskeletal physiological responses of agoraphobics during nonpanic, cognitive, and physiological stress conditions, and how these measures changed during treatment. Marked physiological, particularly autonomic, changes occur during the panic attacks that most of these patients experience. Although all of our 14 patients had a history of panic attacks, we did not specifically attempt to induce them, and no patients panicked during testing. We attempted to answer the question: Do agoraphobic patients have a specific profile of physiological stress responses different from controls?

We expected to find that physiological baseline measures and responses to stressors would change as a function of successful treatment, and that differences would be found between fluorescent and incandescent lighting conditions. We also looked for pretreatment physiological characteristics that would predict clinical treatment response.

LITERATURE REVIEW

Physiological Assessment of Fear, Panic, and Anxiety

A central problem of the field of psychophysiology has been to find specificity in the response patterns to various situations of arousal and anxiety. Lang (1968, 1977) has addressed this problem by postulating the division of response events into three categories: (1) language behavior, (2) motor acts, and (3) changes in the viscera and the level or pattern of muscle tonus. He urged the design of experiments in which measures in all three categories were recorded, with no a priori conclusions as to which behaviors were causal.

Effects of Lighting

There is at least theoretical support for a difference effect between fluorescent and incandescent lighting. Incandescent lights produce *continuous* spectra with some output at all visible wavelengths. Mercury vapor lamps produce *line* spectra, with radiation occurring only at certain wavelengths. Fluorescent lights produce both types, the phosphor producing a continuous radiation and the mercury producting line radiation. Thus the important difference between the two lighting conditions studied here is the presence of spikes of line radiation under fluorescent lighting.

Rapkin (personal communication) has described five patients with agoraphobia, all of whom had panic attacks, who reported that fluorescent lighting aggravated their symptoms. These patients covered the full spcctrum of symptoms seen in agoraphobia, including one with obsessional rituals made worse by fluorescent lighting. All patients responded well to either imipramine or phenelzine.

The small portion of the electromagnetic spectrum with frequencies between about 350 and 750 nanometers is called visible light because humans can recognize its presence cognitively. It is well accepted that this psychophysical phenomenon is mediated via neural connections from the retina through the optic nerve and tract, lateral geniculate body, optic radiations, and finally occipital cortex. However, not all fibers from the retina take this route. Some fibers diverge to descend to the superior colliculus, where they help to mediate visual reflexes such as pupillary size (Heimer 1983). More importantly for our discussion, a newly demonstrated monosynaptic pathway called the retinohypothalamic tract leads from the retina to the suprachiasmatic nuclei of the hypothalamus (Moore and Lenn 1972). This region is thought to play some role in the setting and entrainment of biological clocks (Moore-Ede et al. 1983). The range of light frequencies to which this pathway is sensitive is unknown.

Fluorescent lamps consist of two parts. Inside the glass tube is a low-pressure mercury discharge lamp, which emits almost 90 percent of its energy at a wavelength of 253.7 nm, in the near-ultraviolet range, with other, smaller peaks in the visible range (Sidney and Wolbarsht 1980). The output is modulated in synchrony with the 60-Hz line voltage. Since the lamp discharges on both the positive and negative phases of the cycle, its effective flicker rate is 120 Hz. However, as the lamp ages, the two phases tend to discharge unequally, and so there is a secondary modulation of output at 60 Hz (50 Hz in Europe). This intermittent light energy excites the phosphor coating the inner surface of the tube, which then glows in a continuous manner. These phosphors have enough phosphorescence to smooth the intermittent light output from the mercury discharge. However, the intermittent nature of the output is not completely masked. A slight flicker can be seen by most individuals when viewing the end of a fluorescent lamp via the peripheral retina (Sidney and Wolbarsht 1980). The light output is a combination of intermittent line spectrum radiation from the mercury discharge element and continuous, broad-spectrum radiation from the phosphor. The first practical hot-cathode, low-voltage fluorescent lamp was marketed in 1938. The chief advantage is that of economy. Incandescent lamps, on the other hand, produce a continuous spectrum, with some radiation occurring at all wavelengths and with no sharp frequency peaks. Incandescent lamps do not flicker.

Psychophysiological Studies

Sagberg and Kveim (1981) found that control subjects given single-session biofeedback training of the frontalis electromyograph (EMG) demonstrated significant covariation with levels recorded from other muscle sites about the head and neck, such as the sternocleidomastoid and the masseter. Their work supports the idea that, in most people, such physiological measures rise or fall together with arousal or anxiety.

There are few psychophysiological studies of patients with agoraphobia. Stern and Marks (1973) measured heart rate while subjecting 15 agoraphobic patients to in vivo exposure using four half-hour sessions, or one two-hour session. They found that the greatest decrements in heart rate, as well as in anxiety self-report, occurred during the second hour. Mean heart rate decrements were on the order of five beats per minute (bpm).

Mavissakalian and Schofield (1980) reported preliminary data on self-reported anxiety and on heart rate of three agoraphobic women as they walked a 1.2-km outdoors course, a situation that would presumably stimulate phobic anxiety in these agoraphobics. Assessment was performed before, at midpoint, and after a 12-session behavioral group treatment. In two of the patients, anxiety self-report and heart-rate measures tended to rise and fall together, while in the third patient, who initially improved rapidly but who relapsed after the conclusion of the study, there was marked discordance between the two measures, with heart rate being much higher post-treatment than pretreatment. The study did not appear to control for the exercise effect of the speed of walking itself on heart rate. Again in 1982, Mavissakalian and Michelson monitored heart rate continuously in a group of 26 female agoraphobics while they walked a 0.4-mile behavioral course from the hospital through a congested area. A complete set of analyzable psychophysiological data was obtained for six patients, four of whom were taking imipramine. Heart-rate response (the difference between walking baseline and walking the behavioral course) was 13 beats per minute before treatment, 18 bpm at the fourth week of treatment, and 10 bpm at the end of the 12-week treatment program.

Grunhaus et al. (1983) gave the cold-pressor test to three patients with panic disorder and three patients with agoraphobia with panic attacks, while monitoring changes in prolactin and blood pressure. There were no significant differences between these patients and a control group.

Fredrikson and Ohman (1979) conditioned normal subjects to fear-relevant (snakes and spiders) and fear-irrelevant (flowers and mushrooms) pictures by an electric-shock unconditioned stimulus. Finger-pulse volume and skin-conductance responses were acquired by the conditioned stimuli, and the conditioning to fear-relevant stimuli showed superior resistance to extinction. Heart-rate responses were not acquired by the paradigm.

SUBJECTS

Patients were solicited from clinical referral sources as well as by television and radio publicity. Those who met DSM-III criteria for agoraphobia with or without panic attacks were admitted to the study. To date, 14 such patients have completed the entire psychophysiological protocol, which consists of pretreatment assessment, then treatment with a combination of imipramine and group therapy, followed by post-treatment assessment. Their mean ± SD age was 35 ± 8, and the mean ± SD duration of their symptoms was 10 ± 7 years. There were 13 females and one male in the group. Three patients who met other criteria for inclusion were excluded from this study because they could not remain still in the experimental chair for us to record accurate data.

After assessment, patients were treated with a combination of imipramine and weekly group therapy (see Chapter 2.) They were randomly assigned to either low (100–150 ng/ml) or high (200–250 ng/ml) plasma levels, and dosages were adjusted to maintain these levels throughout treatment. Drug levels were drawn at intervals and measured as the total of imipramine and desipramine by high-performance liquid chromatography.

Control subjects were solicited from student and hospital personnel. They were free of symptoms of panic disorder, agoraphobia, and other major psychiatric illness, and were taking no psychotropic drugs. Controls each had one psychophysiological assessment, while patients were assessed twice, pre and post treatment. The mean ± SD age of control subjects, which was 32.6 yrs ± 8 years, did not differ significantly from that of patients.

PROTOCOL

Conditions

The patients were studied in the Psychophysiology Laboratory of the Department of Behavioral Medicine and Psychiatry at the University of Virginia, Charlottesville. The experimental room was sound attenuated and temperature controlled at 72 F. Incandescent lighting was provided by six flush-mounted reflector units using Westinghouse extended-service 150 W bulbs with rheostat control, shining through Prescolite® diffusers. Fluorescent lighting was provided by four 22-inch Brightstik® units, each 75 W. Both of these light sources were 49 inches above the eye height of a five-foot, 10-inch subject, and we did not control for subject height differences. The incandescent rheostat was set so that each lighting condition produced an equal level of illumination, which we measured as 50 footcandles of light incident upon the patient's head while seated in the experimental chair.

The protocol consisted of 28 two-minute trials, during which subjects were instructed to keep their eyes open. Compliance was monitored by closed-circuit TV. The assessment was divided into four stressor modules. A module consisted of three stress trials, each of which was followed by a rest trial. Each patient received a cognitive and a physiological stressor under each of the two lighting conditions. The cognitive stressor consisted of two minutes of rapid serial subtraction of sevens from 500, followed by a two-minute rest period. This was repeated twice more, for a total of three stress periods and three rest periods. Then a physiological stressor trial was applied for two minutes, three times in succession, also with intervening rest periods. The physiological stressor consisted of a one-minute immersion of the gloved right hand up to the wrist in ice water (modified cold-pressor test). At this midpoint in the protocol, lighting conditions were changed, and the sets of cognitive and physiological stressors were repeated. The type of lighting used first was randomized for the pretest assessment, and for each patient the order was unchanged for posttreatment assessment.

Cognitive-stressor modules always preceded physiological modules.

Measures

Clinical severity and treatment response were measured in part by the use of two five-point scales, the Symptom Severity and Avoidance Behavior Scale (Marks 1969) and the Work and Social Disability Scale (Kelly 1970). The Multiple Affective Adjective Checklist (MAACL) was administered twice during each psychophysiological assessment, once after each lighting condition. We report data from the following psychophysiological measures:

(1) Frontalis EMG activity was measured across a 7 cm interelectrode distance, 2 cm above the eyebrows.

(2) Forearm EMG electrodes were placed by first finding the midpoint of a line between the navicular process of the wrist and the medial humeral condyle, with the hand in supination. One electrode was placed 3.5 cm toward the wrist from this point, and one 3.5 cm toward the elbow, for a total interelectrode distance of 7 cm. Both the frontalis and forearm EMG measures utilized $Ag/AgCl_2$ electrodes and Med Associates electrode gel. The skin was prepared with alcohol and Omniprep such that electrode impedances were kept below 10,000 ohms, as measured by a Grass EZM 3-A impedance meter. Signals were amplified 10,000 times by Med Associates amplifiers, then full-wave rectified and smoothed using a resistance-capacitance network to produce a continuously varying positive signal, with a voltage proportional to EMG activity.

(3) Heart rate was derived from Lead I of the electrocardiogram, which was amplified and beat-to-beat interval determined by a Grass 7P4 EKG-Tachograph. The voltage output of the Tachograph was proportional to the beat-to-beat interval.

(4) Temperature of the left fifth finger was recorded using a Yellow Springs Model 709 thermistor and Med Associates temperature module. The right hand was used for the cold-pressor test.

Each of these four processed signals was subjected to voltage-to-frequency conversion, generating a digital pulse string which was counted and averaged over five-second intervals for each variable using a microcomputer. At the end of each two-minute trial, this data was stored on a floppy disk for later retrieval and analysis.

RESULTS

Clinical Measures

The group mean ± SD pretreatment score on the Work and Social Disability Scale was 4.79 ± 0.43 and fell to 2.07 ± 0.92 after treatment. The Symptom Severity Scale mean ± SD score fell from 4.64 ± 0.5 at pretreatment to 2.07 ± 0.92 at post-treatment. As mentioned, most patients experienced significant clinical improvement.

Patients versus Control Subjects

The control group was compared to the patient group at both pretreatment and post-treatment assessment. The mean heart rate across all trials for the control group was 71.5 bpm, which did not differ significantly from the patient-group mean at pretreatment, which was 74.3 bpm. Mean forearm EMG activity across all trials was 17.5 microvolts (μv) in the patient group at pretreatment, and 6.7 μv in the control subjects, a difference significant at the .02 level using the group *t*-test. At post-treatment, the patient mean was 10.9 μv, still significantly different from the control group. Frontalis EMG mean for the control group was 11.7 μv, significantly different from the pretreatment patient mean of 16.6 μv. At post-treatment, the patient mean of 15.4 μv remained significantly different from that of control subjects. It must be remembered that comparisons of EMG levels between groups are hazardous because of differences in muscle bulk, amount of subcutaneous fat, and skin thickness.

Incandescent versus Fluorescent Lighting Conditions

Control group frontalis EMG levels did not differ significantly between lighting conditions. However, mean ± SD pretreatment frontalis EMG levels in the agoraphobic patients were 16.8 ± 7.1 μv under fluorescent conditions and 15.4 ± 6.0 μv under incandescent conditions. This difference is significant at the $p < .02$ level (paired *t*-test) and showed no significant decrement after treatment. No other measures showed significant lighting effects.

The Multiple Affective Adjective Checklist scores at pretreatment were significantly higher under fluorescent than incandescent lighting in the patients at the $p < .05$ level (paired *t*-test). This difference was not present after treatment. Control subjects did not demonstrate a significant difference between fluorescent and incandescent lighting conditions on this measure. The patients' pre–post changes on this scale did not correlate significantly with pre–post changes on global behavioral outcome measures.

Pretreatment versus Post-treatment

When mean ± SD pretreatment forearm EMG levels were compared with those at post-treatment across all subjects, a drop from 17.5 ± 12.0 μv to 10.9 ± 4.2 μv was seen. This difference is significant at the $p < .05$ level using the paired *t*-test . For each patient the amount of EMG decrease was not correlated with improvement as measured by the Symptom Severity Scale.

Heart rate increased significantly ($p < 0.003$) from 74 ± 8.5 bpm to 89 ± 15.8 bpm during treatment. There was no difference between the low and high imipramine groups on this measure, nor did the amount of heart-rate change correlate significantly with any of the global outcome measures.

Mean ± SD finger temperature fell from 87.6 ±7.2 F to 83.9 ± 8.4 F. This change is of only marginal significance ($p = .084$). Frontalis EMG levels did not change significantly during treatment.

Predictors of Treatment Outcome

Pretreatment measures were regressed upon global outcome change scores to determine to what degree such measures predicted outcome. The mean pretreatment forearm EMG level across all conditions was inversely related to the degree of clinical improvement as measured on the Symptom Severity Scale at the .002 level (linear regression [r] = −.82). That is, lower pretreatment forearm EMGs were associated with better clinical outcomes. In addition, pretreatment heart rate was directly related to clinical improvement on the Work and Social Disability scale at the .04 level (r = +.65). Higher heart rates were associated with better outcomes. The amount of finger temperature change when the opposite hand was placed in cold water was inversely related to the degree of clinical improvement on the Work and Social Disability scale at the .005 level (r = −.77). Smaller temperature changes were associated with better outcomes.

DISCUSSION

The small but significant difference in frontalis EMG levels between the two lighting conditions in patients but not in controls lends some support to the anecdotal reports described above. However, we have not yet studied enough patients with disorders other than agoraphobia under these conditions to know if these findings are limited to this disorder. The other three variables did not demonstrate a significant lighting effect. A more robust lighting effect could have been missed by our experiment in one or more of the following ways. First, the lighting conditions might not have been bright enough. Second, the particular fluorescent lights used were of the self-starting type and were new. Their fluorescent coating makes their flicker less noticeable than in older designs, and they would not exhibit the increase in flickering sometimes seen when older lights age. Third, our study included baseline and stress conditions, but not actual panic conditions. Perhaps if we had attempted more vigorously to induce panic attacks, for example via lactate infusion, we would have seen a greater difference. Fourth, other physiological or psychological measures might have been more sensitive to lighting conditions.

That we did observe an increase in subjectively experienced anxiety in the agoraphobic patients, but not in the controls, under fluorescent lighting is the first experimental support for the anecdotal association between fluorescent lighting and anxiety. It too should be investigated further.

Why did forearm EMG fall as the patients got better, but frontalis EMG remain the same? Earlier psychophysiological work in this area, done mostly on patients with generalized anxiety, tended to view EMG activity as a nonspecific indicator of arousal or anxiety. Our data imply that agoraphobics do not follow this model.

The increase in heart rate during treatment is most likely related to the anticholinergic effects of imipramine. Imipramine has been shown to increase heart rate 10–15 bpm. Since none of our patients panicked during the testing, we do not know if heart-rate increase during panic conditions was affected. Post-treatment heart rate did not differ significantly between low and high imipramine groups. This implies that the anticholinergic effect of the drug, at least on heart rate, is already maximal at the lower plasma level.

Fingertip temperature showed a significant decline during treatment, with an average drop of 5 F measured across all trials. None of our patients had cold extremities as a significant complaint. Cold extremities are usually considered to be a sign of generalized autonomic arousal, but the other physiological measures do not indicate this. Peripheral temperature is largely controlled by vasomotor tone. Since skin and mucosal arterioles demonstrate vasoconstriction to adrenergic stimulation and dilation to cholinergic stimulation (Gilman et al. 1980), the drop in extremity temperature may represent vasoconstriction to the high circulating levels of norepinephrine and epinephrine we observed (see Chapter 2) or anticholinergic effects of the imipramine. In addition, perhaps we need to distinguish between physiological measures of agoraphobics when quiescent and when in their panic state.

Finally, could the increase in heart rate and the decrease in extremity temperature after treatment represent the physiological concomitants of increased generalized arousal? There are two objections to this conclusion. First, these patients were doing much better clinically as measured by global change scores. Second, the anxiety questionnaires given during the psychophysiological assessments showed no change in anxiety from pre- to post-treatment.

Perhaps these unexpected changes indicate that there is something characteristic about the physiological functioning of agoraphobics even when they are not experiencing a panic attack, and which is not directly related to physiological measures of generalized arousal or anxiety. Bruno, Myers, and Glassman (1983) studied cardiovascular measures of autonomic functioning in 22 drug-free unipolar depressed patients and concluded that there was a specific autonomic profile associated with unipolar depression, characterized by decreases in cholinergic and beta-adrenergic activity, as well as an increase in alpha-adrenergic activity. There may exist a characteristic psychophysiological profile for agoraphobics as well. Such specific physiological profiles for specific disorders may, of course, be related merely to the random characteristics of the particular group studied. Replication of these findings will help to answer that question.

SUMMARY

(1) As measured by mean global-outcome scales, this group of 14 agoraphobics showed significant clinical improvement after treatment with imipramine and group therapy.

(2) Frontalis EMG levels under fluorescent lighting conditions were slightly but significantly higher than under incandescent lighting conditions, and this difference persisted during treatment. Subjective anxiety in the agoraphobic patients, but not in controls, increased under fluorescent lighting.

(3) During treatment, mean heart rate increased significantly, probably as a result of the anticholinergic effect of imipramine. Both forearm EMG and finger temperature fell significantly during treatment.

(4) The pretreatment forearm EMG and heart-rate levels and the change in finger temperature during cold-pressor test were significant predictors of treatment outcome as measured by changes in the global behavioral-outcome scales.

(5) The data suggest that there is a specific physiological profile characteristic of agoraphobics, which is dissimilar to physiological profiles of patients with generalized anxiety and general arousal patterns.

References

Barlow DH, Mavissakalian MR, Schofield LD: Patterns of desynchrony in agoraphobia: a preliminary report. Behav Res Ther 18:441–448, 1980

Bruno R, Myers S, Glassman A: A correlational study of cardiovascular autonomic functioning and unipolar depression. Biol Psychiatry 18:227–235, 1983

Fredrikson M, Ohman A: Cardiovascular and electrodermal responses conditioned to fear-relevant stimuli. Psychophysiology 16:1–7, 1979

Gilman AG, Goodman L, Gilman A: Goodman and Gilman's The Pharmacological Basis of Therapeutics. New York, Macmillan, 1980, p. 60

Grunhaus L, Gloger S, Birmacher B, et al: Prolactin response to the cold pressor test in patients with panic attacks. Psychiatry Res 8:171–177, 1983

Heimer L: The Human Brain and Spinal Cord: Functional Neuroanatomy and Dissection Guide. New York, Springer-Verlag, 1983

Kelly D, Guirguis W, Frommer, E, et al: Treatment of phobic states with antidepressants. Br J Psychiatry 116:387–398, 1970

Lang P: Fear reduction and fear behavior: problems in treating a construct. Research in Psychotherapy 3:90–102, 1968

Lang P: Physiological assessment of anxiety and fear, in Behavioral Assessment: New Directions in Clinical Psychology. Edited by Cone JD, Hawkins RP. New York, Brunner/Mazel, 1977

Marks IM: Fears and Phobias. New York, Academic Press, 1969

Marks IM: Agoraphobic syndrome (phobic anxiety state). Arch Gen Psychiatry 23:538–553, 1970

Marks IM, Herst ER: A survey of 1,200 agoraphobics in Britain. Social Psychiatry 5:16–24, 1970

Mavissakalian MR, Schofield LD: Patterns of desynchrony in agoraphobia: a preliminary report. Behav Res Ther 18:441–448, 1980

Mavissakalian M, Michelson L: Patterns of psychophysiological change in the treatment of agoraphobia. Behav Res Ther 20:347–356, 1982

Mayer SE: Neurohumoral transmission and the autonomic nervous system; in The Pharmacological Basis of Therapeutics. Edited by Gilman AG, Goodman LS, Gilman A. New York, Macmillan, 1980

Moore R, Lenn H: A retinohypothalamic projection in the rat. J Comp Neurol 146:1–14, 1972

Moore-Ede M, Czeisler C, Richardson F: Circadian timekeeping in health and disease. Part 1. Basic properties of circadian pacemakers. N Engl J Med 309:469–476, 1983

Sagberg F, Kveim KB: Simultaneous EMGs from six sites during muscular relaxation: a comparison between forehead and forearm feedback. Psychophysiology 18:424-431, 1981

Sidney D, Wolbarsht M: Safety with lasers and other optical sources. New York, Plenum Press, 1980

Stern R, Marks IM: Brief and prolonged flooding: a comparison in agoraphobic patients. Arch Gen Psychiatry 28:270–276, 1973

4

Lactate Induction of Panic: Hypothesized Mechanisms and Recent Findings

Andrew P. Levin, M.D.
Michael R. Liebowitz, M.D.
Abby J. Fyer, M.D.
Jack M. Gorman, M.D.
Donald F. Klein, M.D.

4

Lactate Induction of Panic: Hypothesized Mechanisms and Recent Findings

LACTATE INDUCTION OF PANIC

The association between anxiety and serum lactate was first noted in the context of exercise. Several workers in four different countries observed that anxious patients developed both anxiety symptoms and high blood lactate with physical exercise (Cohen and White 1950; Jones and Mellersh 1946; Linko 1950; Holmgren and Strom 1959). Although anxiety patients are not the only group who develop high lactate levels in the exercise setting, Pitts and McClure (1967) hypothesized that high lactate levels might be anxiogenic.

To test this hypothesis, Pitts and McClure administered intravenous lactate to a group of patients who had histories of long-standing generalized anxiety as well as discrete anxiety attacks. These patients resembled current DSM-III panic disorder classification patients. Thirteen of 14 patients developed panic reactions with 0.5 *M* intravenous (IV) lactate infusion, while only two of the 10 normal controls felt anxious. Serum lactate (DL) was reported to be in the 12–15 m*M* range at the end of the 20-minute infusions in both patients and controls. Pitts and McClure did not measure lactate levels at the actual point of panic, which they report began within one to two minutes of the start of the infusion. Hypertonic glucose of equal molarity to the lactate infusion evoked no response in either group. The patients described their symptoms as similar to those experienced with their "worst attacks." The two normals who developed anxiety symptoms reported headache, dizziness, weakness, and apprehension, stating, "If this is the way nervous people feel I can understand why they get upset." Overall, the patients experienced significantly more numerous and intense anxiety symptoms than controls.

Lactate induction of panic has since been well documented by at least four other groups (Bonn et al. 1971; Fink et al. 1970; Kelly et al. 1971; Knott et al. 1981). In these studies 80–100 percent of patients panicked with infusion of 0.5 *M* lactate, while controls panicked at no more than a 25 percent rate. Investigation into the mechanism of this robust effect, which so nearly approximates in vivo psychopathology, should help elucidate the pathophysiology of spontaneous panic attacks. To date, a variety of mechanisms have been proposed to explain lactate-induced panic attacks (Table 1).

Hypocalcemia Hypothesis

Pitts and McClure initially hypothesized that lactate provoked panic by complexing ionized calcium at the surface of excitable membranes, thereby lowering the threshold for firing. They observed that adding calcium to the lactate infusion considerably diminished the incidence and severity of resulting anxiety symp-

Table 1 Potential Mechanisms for Lactate's Induction of Panic

1. Lowering of ionized calcium
2. Metabolic alkalosis
3. Beta-adrenergic hypersensitivity
4. Peripheral catecholamine release
5. Central noradrenergic stimulation
6. Hyperventilation
7. Endogenous opioid dysregulation
8. Alteration of NAD^+-to-NADH ratio
9. Nonspecific stress

toms (Pitts and McClure 1967). Both patients and raters had no difficulty distinguishing the infusions by observing their effects, with the lactate-plus-calcium infusion consistently falling between the lactate-only and glucose solutions in symptoms produced.

Reviewing this work, Grosz and Farmer (1969) argued that induction of metabolic alkalosis was the more likely precipitant of lactate-induced panic, and further calculated that the lactate dosages administered by Pitts and McClure were insufficient to significantly depress ionized calcium. Pitts himself later found that infusions of ethylenediaminetetraacetic acid (EDTA), a powerful calcium chelator, produced symptoms of tetany in panic patients but did not induce panic attacks (Pitts and Allen 1979).

Alkalosis Hypothesis

Pursuing the alkalosis hypothesis, Grosz and Farmer (1969) infused normals with lactate (DL) to assess changes in serum lactate and bicarbonate during the infusions. At five minutes into the infusion, they found serum lactate (DL) levels to average 3 m*M* and concluded that this level had been reached by Pitts and McClure's patients at the point of panic early in the infusion. Grosz and Farmer also reported bicarbonate levels at this point were approximately 2 m*M* above baseline. This immediate rise in bicarbonate was contrary to Pitts and McClure's contention that metabolic alkalosis occurs after the onset of symptoms; instead, the rise in bicarbonate parallels the onset of symptoms.

Grosz and Farmer then related this result to the naturalistic observation that some anxiety neurotics are prone to involuntary hyperventilation resulting in a respiratory alkalosis. During these episodes patients experience fears of suffocation, danger, and panic-like anxiety. Grosz and Farmer hypothesized that alkalosis of any origin might precipitate panic and went on to infuse 10 normals with separate infusions of bicarbonate and lactate (Grosz and Farmer 1972). These subjects reported an equal number of symptoms such as paresthesias, shakiness, tremor, dizziness, and palpitations with both of the infusions.

While an interesting finding, it does not provide a true test of the Grosz and Farmer hypothesis. To do that, bicarbonate infusions (to produce metabolic alkalosis) or hyperventilation (to produce respiratory alkalosis) would have to be studied in patients vulnerable to spontaneous anxiety attacks. Grosz and Farmer's infusion of bicarbonate into normals does not elucidate the involuntary hyperventilation episodes or panic attacks experienced by anxiety patients with lactate infusion.

Exercise Intolerance

Returning to the original observations, the exact meaning of the exercise–lactate findings is not at all clear. First, during exertion in both normals and anxiety patients, endogenously produced lactate causes a metabolic acidosis, not an alkalosis. Second, abnormal increases in serum lactate occur with exercise in patients with heart disease and individuals with poor exercise tolerance, making the specificity of this phenomenon to anxiety patients highly suspect (Holmgren and Strom 1959). Anxiety patients may refrain from exercise because they fear experiencing the increased sympathetic activity associated with exertion, leading to a secondary exercise intolerance. Systematic biological and behavioral assessment of panic patients during exercise testing is needed to help clarify these issues.

Beta-Adrenergic Hypersensitivity

Pitts himself now believes that patients who experience clinical panic attacks are hypersensitive to beta-adrenergic agonists (epinephrine, isoproterenol) or their metabolic products (lactate) (Pitts and Allen 1979). The similarity between the peripheral effects of sympathomimetics and anxiety prompted researchers as early as the 1930s to infuse epinephrine in anxiety patients. Lindemann and Finesinger (1938) injected epinephrine and mecholyl, a cholinergic agonist, into a series of 20 neurotics with histories of distinct anxiety symptoms along with phobic and depressive syndromes. No controls were employed in this study. Contrary to their expectations, both of these agents produced autonomic arousal with palpitations, headache, and diaphoresis.

With regard to panic, Lindemann and Finesinger found that

each agent caused anxiety symptoms in a particular group of patients. Epinephrine brought on attacks in six patients with histories of worry, anxiety, and depression accompanying their history of "overwhelming attacks of distress." It is unclear if these attacks were similar to our current definition of panic or if this phrase refers to intense rumination as experienced by the one adrenaline responder described in more detail. After adrenaline injection these patients became preoccupied with fear and fragmented in speech and thought. In contrast, the five mecholyl panickers were patients with more "definite phobias and concrete fears" than the adrenaline responders. Following mecholyl injection, the phobics evinced a panic characterized by "alert eagerness to cope with a threatening situation," and maintained good speech contact with the observers. Neither of these groups experienced attacks when infused with the other agent.

A third group of five patients responded to both agents with similar anxiety attacks. This group was felt to have shown "early instability of the autonomic system" and "a trend to hypochondriacal ideas." Lindemann cited the example of a 34-year-old man with a history of asthma and general nervousness punctuated by specific attacks of anxiety. The four patients who did not respond to either of the infusions were labeled as "individuals with more serious personality difficulties." In the discussion section following the paper, Lindemann draws the obvious but often overlooked conclusion that "anxiety is not the same thing in every person who has the complaint." This suggests the need for careful symptom dissection among well-categorized patient groups. A subsequent series of studies during the 1950s failed to derive definitive diagnostic or prognostic categories from the results of epinephrine or mecholyl infusions (Funkenstein et al. 1952; Lunde et al. 1958; Sloane et al. 1958).

In more recent work with adrenergic agonists, results have been equally unenlightening around the issue of panic. Norepinephrine has not been found to produce anxiety (Frankenhauser and Jarpe 1962; Vlachakis et al. 1974). Isoproterenol, a beta-adrenergic agonist, precipitates anxiety in hypertensive patients with a history of mixed anxiety symptoms. Normals and nonanxious hypertensives experience much less anxiety when receiving isoproterenol (Frohlich et al. 1969). These studies lack direct relevance to panic patients and to the lactate infusion per se.

There is good evidence that peripheral epinephrine secreted from the adrenal medulla rises dramatically with the stress of public speaking. In a study of house officers, Dimsdale and Moss (1980) found that plasma epinephrine more than doubled from a prespeaking mean of 117 pg/ml to 406 mg/ml during the initial moments of public speaking. This peripheral rise in epinephrine represents a response to the stress of speaking and is associated with a rapid increase in heart rate. Public speaking or performance anxiety can be partially attenuated with pretreatment by beta blockers, as demonstrated in the work of Neftel et al. (1982) with stage fright in musicians. However, the analogy between stage fright and panic attacks is not substantiated by research findings. Acute beta blockade with propranolol does not block lactate-induced panic attacks (Gorman et al. 1983). Further, as reported in the following section, we do not find dramatic increases in peripheral epinephrine accompanying lactate-induced panic attacks.

Central Noradrenergic Mechanisms

Presumably based on central adrenergic mechanisms, yohimbine, an alpha-adrenergic antagonist, has been shown to cause autonomic arousal and marked emotional distress (Garfield et al. 1967; Holmberg and Gershon 1961). Holmberg and Gershon (1961) found that yohimbine injection in a heterogeneous population of patients and controls uniformly brought on flushing, perspiration, salivation, pupillary dilation, and a rise in heart rate and blood pressure followed by a restless, tremulous, irritable state which they labeled anxiety. In a later paper, Garfield et al. (1967) compared yohimbine injection to epinephrine injection in "schizophrenic" and "nonschizophrenic" patients, concluding that the combination of autonomic and psychic symptoms induced by yohimbine more closely approximated clinical anxiety states than those precipitated by epinephrine infusion. This conclusion cannot be extended to panic itself because pretreatment with imipramine, a clinically effective antipanic agent (Klein 1981) augmented the yohimbine response (rather than attenuated it) in Gershon's earlier work (Holmberg and Gershon 1961).

Although it is therefore difficult to generalize from these results to panic patients, yohimbine's action is of particular interest because it is felt to act centrally at the locus coeruleus. This nucleus is the major source of noradrenergic innervation for higher centers and also appears to be involved in opiate withdrawal (Cooper et al. 1982; Gold et al. 1979), a state characterized by some anxiety symptoms (e.g., anxiety, increased heart rate, and restlessness). These observations have led to speculations that panic states may result from central noradrenergic or opioid dysregulation (Klein 1981). We will consider these hypotheses more carefully in light of our recent findings.

Additional Hypotheses

Other hypotheses advanced to explain lactate's effects include alteration of the ratio of oxidized to reduced nicotinamide adenine dinucleotide (NAD/NADH), a function of lactate-to-pyruvate conversion, and nonspecific stress (Ackerman and Sachar 1974). The latter is well articulated by Ackerman and Sachar. They propose a conditioned phobic response model in which susceptible patients respond with panic attacks to any of several somatic symptoms that have come to be associated with anxiety. This hypothesis most directly attends to the implicit question of the relationship between peripheral effects of lactate and the phenomenon of anxiety, a psychic state of central origin.

Arguing against this is the reported failure of EDTA (Pitts and Allen 1979), mental arithmetic (Kelly et al. 1971), or the cold-pressor test (Grunhaus et al. 1983) to provoke panic in clinically vulnerable patients, despite the prominent physiological changes evoked by these challenges. The testing of other agents that have some but not all of the biological effects of racemic sodium lactate may be one way to assess lactate's specificity in precipitating panic.

CURRENT RESEARCH

To delineate the mechanism by which lactate induces panic attacks, a multidisciplinary longitudinal investigation of lactate provocation was initiated. Patients who met DSM-III criteria for panic disorder or agoraphobia with panic attacks were prospectively entered into infusion and treatment protocols. Implications of our findings for understanding how lactate provokes panic attacks are summarized below.

Procedures and subject selection of the lactate study have been detailed elsewhere (Appleby et al. 1981). In brief, patients who met the criteria, as well as normal controls, underwent adaptation and physiological monitoring on the first day, returning to the laboratory the next day when intravenous lines were started and 5 percent dextrose in water was administered slowly for 28 minutes and then rapidly for two minutes. Following this baseline period, 0.5 *M* racemic sodium lactate, 10 ml/kg body weight was administered over a 20-minute period, unless the patient panicked, at which point the lactate infusion was stopped. On occasion, infusions took longer than 20 minutes to administer, and infusions lasting up to 22 minutes were included in our data analysis. Sequence and duration of infusions were known to staff but not to the subjects. Written informed consent was obtained from each patient after detailed explanation of all study procedures.

In judging that a patient panicked, we required both a crescendo of extreme apprehension or fear and the DSM-III physical symptoms for a panic attack—tachycardia, shortness of breath, sweating, faintness, etc. Physical symptoms alone did not suffice; nor did we score panic if a patient reported that he would be frightened but for the presence of a physician at the infusion. Other studies rated this latter group as panickers (Bonn et al. 1971; Kelly et al. 1971; Pitts and McClure 1967).

FINDINGS AND REASSESSMENT OF MAJOR HYPOTHESES

Pitts and McClure's finding that adding calcium to lactate infusions lowered the incidence of panic (Pitts and McClure 1967) is compatible with, but does not prove, a panicogenic role for hypocalcemia. Adding calcium could simply have offset the effect of lactate on a noncalcium system. Our preliminary findings are that panickers, at the point of panic, do not have lower ionized-calcium levels than comparably timed nonpanicking patients. This suggests that hypocalcemia per se does not cause panic attacks in clinically vulnerable patients. It is possible, however, that lowered calcium, in interaction with some other variable, is panicogenic.

Patients who panic with lactate also did not develop greater pH elevation than nonpanicking patients. Apparently no absolute pH threshold exists beyond which clinically vulnerable patients automatically panic. Also arguing against alkalosis as the mechanism for lactate panic is our unpublished finding that room air hyperventilation does not regularly provoke panic in patients who panicked with lactate (Gorman, personal communication). Panic patients do demonstrate greater minute ventilation during exercise than normals, but they do not panic consistently under these conditions (Cohen and White 1950).

Our catecholamine findings suggest that lactate-induced panic attacks are biochemically different from performance anxiety. Although Dimsdale and Moss (1980) found that public-speaking anxiety regularly was accompanied by twofold or greater plasma epinephrine increase, we found no evidence for rapid increase in plasma epinephrine as the cause of lactate-induced attacks. This could explain why acute beta-adrenergic blockade may ameliorate public-speaking anxiety, but was found ineffective in blocking lactate panic (Gorman et al. 1983). Our findings do not exclude predisposing or pathoplastic roles for epinephrine. Our finding of baseline-heightened arousal among panickers prior to lactate infusion lends some credence to epinephrine mediating a predisposition to panic.

Elevated norepinephrine was found to accompany lactate panic on only half of our measurement periods. We feel this was more likely a reflection than a cause of panic, because elevated plasma norepinephrine is due to spillover into plasma from noradrenergic synapses and presumably reflects heightened peripheral sympathetic activity.

As noted, Pitts argued that panic patients are hypersensitive to all beta-adrenergic agonists. Arguing against a direct beta-adrenergic mechanism for lactate induction of panic is our recently reported finding that pretreatment of patients with 0.2 mg/kg of intravenous propranolol did not prevent lactate-induced panic despite evidence of peripheral beta-adrenergic blockage (Gorman et al. 1983).

The point of panic changes seen in the lactate infusion study do, however, include both elevated heart rate and variable increase in blood pressure. These most likely reflect the result of central sympathetic activation (Holmberg and Gershon 1961). The locus coeruleus (the principal noradrenergic nucleus in the CNS) and its projections have already been suggested as the mediator of fear response in primates (Klein 1981) and in rodents (Gray 1978). That they play a role in human panic attacks as well is suggested by the findings of increased plasma MHPG in phobic anxious patients during in vivo exposure (Ko et al. 1983) and by two studies in which oral clonidine, a locus coeruleus inhibitor, was administered to panic patients on a chronic basis and found to at least transiently block panic attacks (Hoehn-Saric 1981; Liebowitz et al. 1981). In addition, stress has been shown to increase electrophysiological activity in single-unit locus coeruleus recordings in live animals (Cooper et al. 1982).

A role for central noradrenergic pathways as a specific substrate for panic is, however, still speculative. Clonidine may not be specific for noradrenergic systems, since it reverses the non-adrenergically mediated rhinorrhea and lacrimation attendant to narcotic withdrawal (Gold et al. 1979). The effects of clonidine overdose can be reversed by naloxone (Qulig et al. 1982). Locus coeruleus activity in animals is enhanced by a variety of non-anxiety sensory stimuli, suggesting that this noradrenergic nucleus serves in a global orienting manner rather than as a specific neurochemical substrate for anxiety (Cooper et al. 1982). In addition, while heart-rate increase appeared consistently to accompany lactate panic, we did not see the consistent elevations of blood pressure that would be expected to accompany locus coeruleus discharge.

Lactate infusions may also cause panic by alteration of the NAD^+ to NADH ratio. NAD^+, the oxidized form of nicotinamide adenine dinucleotide, is a necessary coenzyme for the oxidative metabolism of carbohydrates, lipids, and amino acids (Orten and Newhaus 1982). In the absence of oxygen (anaerobic conditions), the conversion of pyruvate to lactate helps replenish the body's store of NAD^+ (Figure 1). What occurs during lactate infusion is a huge increase in lactate that, contrary to normal physiology, has not been generated from pyruvate. The NAD^+ to NADH ratio is thus temporarily shifted in favor of NADH.

Whether the shift in the NAD^+ to NADH ratio or other aspects of lactate metabolism cause panic are being tested by the administration of D-lactate and pyruvate. Standard lactate infusions contain an equal mix of L and D isomers of lactate. Pyruvate infusions would have the opposite effects on the NAD^+ to NADH reaction. However, only L-lactate can be converted to pyruvate. If D-lactate or pyruvate proves panicogenic, then the fall in NAD^+ relative to NADH will be excluded as a possible mechanism. If D-lactate does not produce panic, the specific effects of lactate metabolism will require further study. Our current findings of no differences in lactate and inconsistent differences in pyruvate when measured at the point of panic do not settle this question.

Of special interest is the possibility that the metabolism of lactate gives rise to transient elevations of cerebral CO_2 levels, even though overall body P_{CO_2} falls during a lactate infusion because of

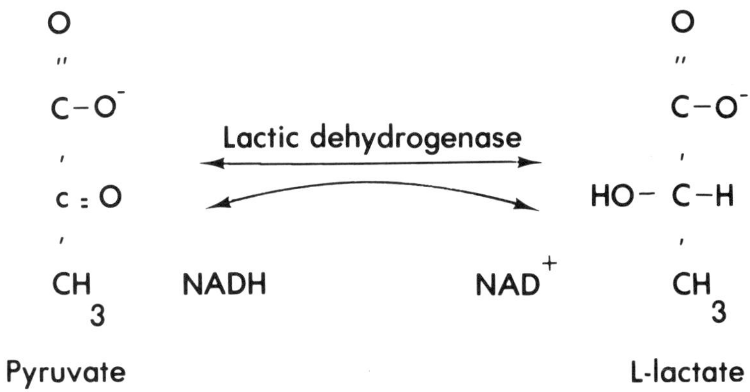

Figure 1. Conversion of lactate to pyruvate with the reduction of NAD^+.

hyperventilation. This could provide a common mechanism for the panicogenic effects of lactate and carbon dioxide. Recently, Gorman et al. have found that inhalation of 5 percent CO_2 produces panic in clinically vulnerable patients, reproducing in a controlled fashion the earlier finding that rebreathing (which produced a CO_2 concentration of 4 percent) precipitated anxiety symptoms in a majority of patients with a history of anxiety attacks (Cohen and White 1950). Elevated cerebral CO_2 is a locus coeruleus stimulant (Svensson et al. 1978) and also powerfully stimulates the brain respiratory center (see Chapter 5). Also linking lactate and CO_2 is their common effect in increasing cerebral blood flow (Mathew et al. 1982).

To summarize, our data do not support the depression of ionized calcium or induction of metabolic alkalosis as sufficient triggers to cause panic during lactate infusions. Whether either or both serve to cause panic attacks in already aroused patients, however, remains to be elucidated. A peripheral catecholamine surge also does not appear to be the mechanism by which lactate causes panic, although epinephrine may play a predisposing role. Evidence of heightened central noradrenergic activity is present during many but not all panic attacks; the significance of this remains to be established. Further studies of hyperventilation, CO_2, the metabolic effects of lactate, and the nonspecific-stress hypothesis are needed.

Overall, our data suggest that no peripheral biochemical or physiological change serves as a trigger for lactate panic. Rather, lactate-induced panic attacks appear to be primarily central nervous system phenomena with inconsistent peripheral sequelae.

References

Ackerman SH, Sachar EJ: The lactate theory of anxiety: a review and reevaluation. Psychosom Med 36:69–81, 1974

Appleby IL, Klein DF, Sachar EJ, et al: Biochemical indices of lactate-induced panic: a preliminary report, in Anxiety, New Research and Changing Concepts. Edited by Klein DF, Rabkin J. New York, Raven Press, 1981, pp 411–430

Bonn JA, Harrison J, Rees WL: Lactate-induced anxiety: therapeutic applications. Br J Psychiatry 119:468–471, 1971

Cohen ME, White PD: Life situations, emotions and neurocirculatory asthenia (anxiety neurosis, neurasthenia effort syndrome). Transactions of the Association for Research in Nervous and Mental Diseases 29:832–869, 1950

Cooper JR, Bloom FE, Roth RH: The Biochemical Basis of Neuropharmacology. New York, Oxford University Press, 1982

Dimsdale JE, Moss J: Short-term catecholamine response to psychological stress. Psychosom Med 42:493–497, 1980

Fink M, Taylor MA, Volavka J: Anxiety precipitated by lactate. N Engl J Med 281:1129, 1970

Frankenhauser M, Jarpe G: Psychophysiological reactions to infusion of a mixture of adrenaline and noradrenaline. Scand J Psychol 3:21–29, 1962

Frohlich ED, Tarazi RC, Dustan HP: Hyperdynamic beta-adrenergic circulatory state. Arch Intern Med 123:1–7, 1969

Funkenstein DH, Greenblatt M, Solomon HG: An automatic nervous system test of prognostic significance in relation to electroshock treatment. Psychosom Med 14:347–362, 1952

Garfield SL, Gershon S, Sletten I, et al: Chemically induced anxiety. International Journal of Neuropsychiatry 3:426–433, 1967

Gold MS, Redmond DE, Kleber HD: Noradrenergic hyperactivity in opiate withdrawal supported by clonidine reversal of opiate withdrawal. Am J Psychiatry 136:100–102, 1979

Gorman JM, Levy GF, Liebowitz MR, et al: Effect of acute adrenergic blockade on lactate-induced panic. Arch Gen Psychiatry 40:1079–1082, 1983

Gray JA: The neuropsychology of anxiety. Br J Psychol 69:417–434, 1978

Grosz HJ, Farmer BB: Blood lactate in the development of anxiety symptoms: a critical examination of Pitts and McClure's hypothesis and experimental study. Arch Gen Psychiatry 21:611–619, 1969

Grosz HJ, Farmer BB: Pitts and McClure's lactate-anxiety study revisited. Br J Psychiatry 120:415–418, 1972

Grunhaus L, Gioger B, Birmacher B, et al: Prolactin response to cold pressor test in patients with panic attacks. Psychiatry Res 8:171–177, 1983

Hoehn-Saric R, Merchant AF, Keyser ML, et al: Effects of clonidine on anxiety disorders. Arch Gen Psychiatry 38:1278–1282, 1981

Holmberg G, Gershon S: Autonomic and psychic effects of yohimbine hydrochloride. Psychopharmacologia 2:92–106, 1961

Holmgren A, Strom G: Blood lactate concentration in relation to absolute and relative work load in normal men, and in mitral stenosis, atrial septal defect, and vasoregulatory asthenia. Acta Med Scand 163:185–193, 1959

Jones M, Mellersh V: Comparision of exercise response in anxiety states and normal controls. Psychosom Med 8:180–187, 1946

Kelly D, Mitchell-Heggs N, Sherman D: Anxiety and the effects of sodium lactate assessed clinically and physiologically. Br J Psychiatry 119:129–141, 1971

Klein DF: Anxiety reconceptualized, in Anxiety, New Research and Changing Concepts. Edited by Klein DF, Rabkin J. New York, Raven Press, 1981, pp 235–263

Knott V, Chaudry R, Lapierre YD: Panic induced by sodium lactate: electrophysiological correlates. Prog Neuropsychopharmacol 5:511–514, 1981

Ko GN, Elsworth JD, Roth RH, et al: Panic-induced elevation of plasma MHPG levels in phobic-anxious patients. Arch Gen Psychiatry 40:425–430, 1983

Liebowitz MR, Fyer AJ, McGrath P, et al: Clonidine treatment of panic disorder. Psychopharmacol Bull 3:122–124, 1981

Lindemann E, Finesinger JE: The effect of adrenalin and mecholyl in states of anxiety in psychoneurotic patients. Am J Psychiatry Sept: 353–370, 1938

Linko E: Lactic acid response to muscular exercise in neurocirculatory asthenia. Annales de Medecine Interne Fenniae 39:161–176, 1950

Lunde F, Mansfield E, Smith JA: Mecholyl chloride as a prognostic test in psychiatric patients. J Nerv Ment Dis 127:430–436, 1958

Mathew RJ, Weinman ML, Claghorn JL: Anxiety and cerebral blood flow, in The Biology of Anxiety. Edited by Mathew RJ. New York, Brunner/Mazel, 1982, pp 23–36

Neftel KA, Adler RH, Kappeli L, et al: Stage fright in musicians: a model illustrating the effect of beta blockers. Psychosom Med 44:461–469, 1982

Orten JM, Neuhaus OW: Human Biochemistry. St. Louis, CV Mosby Co, 1982

Pitts FN Jr, Allen RE: Biochemical induction of anxiety, in Phenomenology and Treatment of Anxiety. Edited by Fann WE, Karacan I, Pokorny AD, Williams RL. New York, Spectrum Publications, 1979, pp 125–140

Pitts FN Jr, McClure JN Jr: Lactate metabolism in anxiety neurosis. N Engl J Med 277:1328–1336, 1967

Qulig K, Duffy J, Rumack DH: Naloxone for treatment of clonidine overdose. Letter to the editor. JAMA 247:1697, 1982

Sloane RB, Kewus DJ, Slater P: Prognostic value of adrenaline and mecholyl response to electroconvulsive therapy. J Psychosom Res 2:271–273, 1958

Svensson TH, Persson R, Wallin L: Anxiolytic action of clonidine. Nordisk Psykiatrisk Tidsskrift 32:439–441, 1978

Vlachakis ND, DeGuia D, Mendelowitz M, et al: Hypertension and anxiety, a trial with epinephrine and norepinephrine infusion. Mt Sinai J Med 41:1615–1625, 1974

5

Evidence That Panic Disorder Has a Metabolic Cause

Daniel B. Carr, M.D.
David V. Sheehan, M.B., B.Ch.

5

Evidence That Panic Disorder Has a Metabolic Cause

Panic disorder seems to be a distinct syndrome, different from the common response-to-stress anxiety of normal humans ("adjustment disorder with anxious mood" in the DSM-III). It has a familial occurrence (Crowe et al. 1980) and has a distinctive age of onset and sex distribution (Sheehan et al. 1981). It does not remit during benzodiazepine or phenothiazine therapy, but improves with MAO-inhibitor and tricyclic-antidepressant treatments (Klein 1968; Sheehan et al. 1980). Panic is reliably evoked in affected individuals but not in normals during infusions of sodium lactate (Pitts and McClure 1967). This suggests that panic disorder is a biological disease (Sheehan 1982).

Herein we will outline the common metabolic features of lactate infusions and carbon dioxide exposure—both used for evoking symptoms of panic under laboratory circumstances in individuals with panic disorder. Our aim will be to outline a heuristic model of this disorder, that is, one that seeks to integrate the available information in a provisional yet testable fashion (Carr and Sheehan, in press). Intriguingly, in many respects the most parsimonious model is one in which panic disorder reflects a dysfunction of brain-stem arousal centers ordinarily activated in normals only by metabolic "danger signals."

LACTATE AND CARBON DIOXIDE: EQUIVALENT SIGNALS

Lactic acid, formed by reduction of pyruvate, is an end point in anaerobic carbohydrate metabolism (Cohen 1976). This reaction requires the cofactor NADH:

$$\text{pyruvate} + \text{NADH} + H^+ = \text{lactate} + \text{NAD}^+$$

By the law of mass action,

$$\frac{[\text{lactate}]}{[\text{pyruvate}]} = K_{LDH} \times \frac{[\text{NADH}]}{[\text{NAD}+]} \times [H^+].$$

Therefore rises in $[H^+]$— i.e, decreased pH—are in a fundamental way equivalent to an increases in the lactate/pyruvate ratio since both increase the numerical value of this equation.

In 1967 Pitts and McClure published a double-blind study showing that infusions of 0.5 M sodium lactate evoked anxiety in susceptible patients but not in normals. These findings prompted Fink to suggest that antianxiety agents be tested by this technique (Fink et al. 1969). Unblinded studies by Kelly et al. (1971) and Appleby et al. (1981) suggested that clinical anxiolytic effectiveness of MAO inhibitors or imipramine correlated with a diminution of symptoms during a rechallenge with lactate. We are now witnessing similar trends in a double-blind, placebo-controlled study currently in progress, in which chronic treatment with a triazolobenzodiazepine, alprazolam, seems to produce both outpatient clinical improvement as well as a blunting of lactate-induced anxiety (Carr and Sheehan, in press).

Such observations, derived from studies of distinct classes of drugs, point to a final common pathway linking spontaneous activity of the disease and vulnerability to lactate. Is lactate simply a nonspecific stressor? This does not seem to be the case, since normals do not complain that the lactate infusion is noxious, nor do susceptible patients panic when infused under identical conditions with placebo solutions of equal volume and osmolarity (DB Carr, DV Sheehan, OW Surman, et al., unpublished observations), nor has anxiety been reported to follow the lowering of calcium during EDTA infusion in patients with panic disorder (Pitts, personal communication).

Since the behavioral effects of lactate must ultimately involve a central action, it is noteworthy that lactic acid is almost entirely ionized at physiological pH, and for this reason, as well as its hydrophilicity, little might be predicted to cross the blood-brain barrier (Weyne and Leusen 1974; Leusen and Weyne 1975). However, in animals there are stereospecific, saturable facilitated-transport systems for L-lactate from blood to brain (Oldendorf 1972; Nemoto and Severinghaus 1974; Nemoto et al. 1974). In addition, several brain regions lack a blood-brain barrier (Martin et al. 1977), amongst them the ventral medullary surface, which serves as a central chemoceptor area or biological "smoke alarm" responding to hypercarbia or acidosis (Dermietzel 1975).

Cerebrospinal fluid lactate concentrations rise only modestly during intravenous lactate infusions sufficient to raise the blood level to 10 m*M* or more in animals (Weyne and Leusen 1974; Leusen and Weyne 1975). For this reason the anxiety generated during brief (about 20 minutes) lactate challenges is not likely to be the result of widespread entry of lactate into the brain, but instead is the result of a combination of rapid, widespread effects of systemic alkalosis (e.g. cerebral ischemia) plus passive elevations of the lactate–pyruvate ratio in restricted regions such as the chemoceptors. Tissue hypoperfusion or pharmacologic elevation of the lactate–pyruvate ratio should both serve to raise intracellular H^+ in medullary chemoceptors.

Our hypothesis that increases in the lactate–pyruvate ratio and concomitant decreases in chemoceptor intraneuronal pH are linked to the clinical experience of anxiety is readily testable, since according to it any maneuver that yields these biochemical changes should produce anxiety in susceptible patients. Overlapping lines of evidence already available suggest this to be so. For example, brain surface pH drops promptly during inhalation of 5 percent carbon dioxide in air (Ahmad et al. 1975). It has been known since World War I that these patients are intolerant to wearing gas masks due to their inability to tolerate rebreathing of carbon dioxide (Cohen and White 1972). At the present time, exposure to 5 percent carbon dioxide in air is employed as an alternative to lactate infusion as a provocative maneuver to induce anxiety in susceptible patients (Griez and van den Hout 1983).

This model is unconventional, yet the word *anxiety* has its origin in the Latin *angere*, meaning to choke or throttle. Were we to recast this primal insight into contemporary terms, then, we would summarize by saying that the two most widely used maneuvers to induce anxiety in patients with panic disorder appear to share a final common metabolic path in which intraneuronal pH is promptly lowered within permeable brain regions, which in turn leads to arousal by well-defined neuronal pathways.

Although a defective cellular "smoke alarm" is an economical explanation, it is not a precise one since numerous cellular regulatory processes are closely integrated. These include the regulation of intracellular pH, the tonic and phasic functioning of multiple ion pumps and channels, the phosphorylation state of the cell (i.e., [ATP]/[ADP]), and the behavior of the calcium–calmodulin complex. We have presented arguments at length elsewhere (Carr and Sheehan, in press) that lend weight to the idea that dysfunction of a minor ion channel could well account for the pattern of inheritance (Pauls et al. 1980) and metabolic alterations presumed to occur in panic disorder. Of course, abundant descending inputs are capable of modulating ventilatory drive or arousal (Kalia 1977; St. John 1977), but for purposes of our model one need only posit a dysregulation of neural pH, redox state, or associated function in the ventral medulla or a nearby efferent projection. The possibility that a projection from the central chemoceptor may be dysfunctional corresponds to the image of an abnormally high gain of a signal that is itself appropriate. Since the syndrome of panic disorder will in the future undoubtedly be recognized to include subtypes, such as the 15 to 20 percent of patients who meet DSM-III criteria for the disorder but who fail to respond to lactate or carbon dioxide, one must keep an open mind that certain patients may have dysfunctional sensors and others may have inappropriate amplification.

CLINICAL ASPECTS OF VENTILATION–AROUSAL

Regardless of the precise nature of the subcellular "lesion" in panic disorder, any means to enhance sensitivity of central chemoceptors (or to augment the response to the activation of these areas) should have the effect of worsening the disorder. Two types of drugs are of clinical use as central ventilatory stimulants: methylxanthines (e.g., caffeine) and progestins (e.g., progesterone) (Goodman and Gilman 1975). Caffeine is well known to be anxiogenic in patients with panic disorder. Progesterone challenges have not yet been attempted in patients with panic disorder, but this hormone is normally secreted in the luteal phase of the menstrual cycle (Martin et al. 1977), and evidence has appeared that links the premenstrual secretion of progesterone with significant falls in end-tidal carbon dioxide and blood-base excess shortly before menses in normal women (England and Farhi 1976). According to our model, such enhancement of central respiratory drive by changes in progesterone during the luteal phase might underlie clinical observations of worsened anxiety during this time in susceptible patients. Further, the typical age of onset of panic disorder in women, early reproductive life, may simply reflect the priming of respiratory drive, by progesterone serving as a "promoter" in a genetically induced disease. Interestingly, ventilatory responsiveness to CO_2 shows a significant correlation with extraversion scores in women (Skatrud et al. 1978). We are now examining ventilatory responsiveness to CO_2 in patients with panic disorder before and after successful therapy, an approach anticipated years ago by Bulow (1963), who quantitated CO_2 sensitivity and simultaneous EEG measures of arousal in normal subjects.

By linking arousal and anxiety with the stimulation of central chemoceptors, it may be possible to render moot controversy as to whether hyperventilation syndrome (Magarian 1982) is distinct from panic disorder, if they in fact share common pathogenetic features. The tendency for panic patients to display biochemical features commonly identified with the hyperventilation syndrome, such as low arterial CO_2 and low serum phosphate, further supports this viewpoint.

RELATION TO OTHER ANXIETY MODELS

Stimulation of the locus coeruleus in animals produces behavior that seems to correspond to anxiety, an effect that is blocked by lesions of this site (Redmond and Huang 1979). Clonidine-induced reduction of central norepinephrine release lessens the severity of opiate- or alcohol-withdrawal symptoms and in animals reduces the firing rate of neurons within the locus coeruleus. Plasma MHPG levels, employed by many (Charney et al. 1982) but not all (Kopin et al. 1983) workers as an estimate of central noradrenergic activity, correlate with anxiety in phobic anxious patients and fall during therapy with imipramine or clonidine in such patients (Ko et al. 1983). Clonidine, however, appears to be less effective than other anxiolytic agents in the treatment of outpatients with panic anxiety or generalized anxiety, and may worsen anxiety (Hoehn-Saric et al. 1981). Moreover, the positive effects of clonidine are not specific to one subtype of anxiety (i.e., either panic or generalized) (Hoehn-Saric et al. 1981). Further, in humans given yohimbine (an alpha-2 antagonist expected to enhance central norepinephrine release), the resulting anxiety is abolished by diazepam pretreatment but diazepam has no effect upon yohimbine-induced rises in plasma MHPG (Charney et al. 1983). Also, beta-adrenergic blockade is only partially effective in improving panic disorder or lactate-induced symptoms (Gorman et al. 1983).

The implications of these results—namely, that noradrenergic systems underlie some but not all aspects of anxiety, and that their activation is not specific to panic disorder—are consistent with our own negative preliminary results on plasma MHPG levels during lactate infusions in an ongoing study in collaboration with Dr. George Heninger. So, our own preliminary findings to date, and the literature reviewed above, suggest that noradrenergic pathways may be mobilized in, but are neither specific nor necessary for, panic anxiety. This conclusion regarding the nonprimacy of noradrenergic systems in panic disorder fits well with our model, for central chemoceptor activation is a potent stimulator of central and peripheral sympathetic activity.

Why are the normally anxiolytic benzodiazepines ineffective in panic disorder? If, as we surmise, panic disorder corresponds to malfunction of medullary chemoceptive areas or their nearby efferent zones, then we need not demand that rostral "downstream" inhibitory systems such as the GABA neurons be mobilized to produce remission. Antidepressants themselves are well known to modulate medullary neuronal excitability; indeed, their efficacy in other clinical settings such as chronic pain is felt to derive in large part from their actions upon monoaminergic, particularly serotonin, neurons in the medulla (Carr and Carr 1983). Therefore it is plausible that antidepressants with predominant effects upon serotonin metabolism (e.g., zimelidine), or which contain a triazolo group without a benzodiazepine moiety (e.g., trazodone) could be beneficial in panic anxiety. It would further appear plausible that given the widespread distribution of rostral projections from the medulla, the diffuse excitatory effects of medullary activation could be mimicked by impeding GABA-ergic inhibitory tone at the target sites of such activation, as evidently is the case when severe anxiety is induced by administration of a beta-carboline (Ninan et al. 1982; Dorow et al. 1983).

As to the apparent benefit (in our pilot studies) of the benzodiazepine alprazolam in the treatment of panic disorder, we note that this compound is not simply a benzodiazepine, but rather is a triazolobenzodiazepine, and so might well have pharmacologic effects beyond those of conventional benzodiazepines. Therefore we would surmise that the efficacy of alprazolam and related triazolobenzodiazepines in panic may be due to the action of its triazolo moiety upon monoamine, especially serotoninergic, neurons, while the benzodiazepine moiety serves to reduce the anticipatory anxiety often present in long-standing cases of panic disorder.

We have sketched a model that is an initial attempt to take the clinical phenomena of panic disorder and to integrate them within a provisional but testable framework. In this sense our model may be called heuristic, in contrast to what may be termed an algorithmic approach in which the clinical phenomena of panic disorder are forced into one or another pre-existing scheme.

It is likely that this syndrome will evolve in similar fashion to other diseases once felt to be psychiatric and later viewed as neuroendocrinologic or metabolic as their pathophysiology was clarified. If so, panic anxiety, now conceptualized in cognitive terms, may one day be viewed simply as but one consequence of a primary brain-stem neuronal dysregulation.

References

Ahmad HR, Berndt J, Loeschcke HH: Bicarbonate exchange between blood, brain extracellular fluid and brain cells at maintained pCO_2, in Acid–Base Homeostasis of the Brain Extracellular Fluid and the Respiratory Control System. Edited by Loeschcke HH. Littleton, Mass, Thieme Edition/Publishing Sciences Group, 1975

Appleby IL, Klein DF, Sachar EJ, et al: Biochemical indices of lactate-induced panic: a preliminary report, in Anxiety: New Research and Changing Concepts. Edited by Klein DF, Rabkin J. New York, Raven Press, 1981

Bulow K: Respiration and wakefulness in man. Acta Physiol Scand [Suppl] 209:1–110, 1963

Carr DB, Carr JM: Role of brain opiates in pain relief, in Bone Metastasis. Edited by Stoll BA, Parbhoo S. New York, Raven Press, 1983

Carr DB, Sheehan DV: Panic anxiety: a new biological model. J Clin Psychiatry (in press)

Charney DS, Heninger GR, Sternberg DE: Assessment of alpha2 adrenergic autoreceptor function in humans: effects of oral yohimbine. Life Sci 30:2033–2041, 1982

Charney DS, Heninger GR, Redmond DE Jr: Yohimbine induced anxiety and increased noradrenergic function in humans: effects of diazepam and clonidine. Life Sci 33:19–29, 1983

Cohen ME, White PD: Neurocirculatory asthenia: 1972 concept. Milit Med 137:142–144, 1972

Cohen RD: Disorders of lactic acid metabolism. Clin Endocrinol Metab 5:613–625, 1976

Crowe RR, Pauls DL, Slymen DI, et al: A genetic study of anxiety neurosis. Arch Gen Psychiatry 37:77–79, 1980

Dermietzel R: Central chemosensitivity, morphological studies. In Acid Base Homeostasis of the Brain Extracellular Fluid and the Respiratory Control System. Edited by Loeschcke HH. Littleton, Mass, Thieme Edition/Publishing Sciences Group, 1975

Dorow R, Horowski R, Paschelke G, et al: Severe anxiety induced by FG 7142, a beta-carboline ligand for benzodiazepine receptors. Lancet 2:98–99, 1983

England SJ, Farhi LE: Fluctuations in alveolar CO_2 and in base excess during the menstrual cycle. Respir Physiol 26:157–161, 1976

Fink M, Taylor MA, Volavka J: Anxiety precipitated by lactate. N Engl J Med 281:1429, 1969

Gilman AG, Goodman LS, Gilman A: The Pharmacological Basis of Therapeutics, 6th ed. New York, Macmillan, 1980

Gorman JM, Levy GF, Liebowitz MR, et al: Effect of acute beta-adrenergic blockade on lactate-induced panic. Arch Gen Psychiatry 40:1079–1082, 1983

Griez E, van den Hout MA: Treatment of phobophobia by exposure to CO_2-induced anxiety symptoms. J Nerv Ment Dis 171:506–508, 1983

Hoehn-Saric, R, Merchant AF, Keyser ML, et al: Effects of clonidine on anxiety disorders. Arch Gen Psychiatry 38:1278–1282, 1981

Kalia M: Neuroanatomical organization of the respiratory centers. Fed Proc 36:2405–2411, 1977

Kelly D, Mitchell-Heggs N, Sherman D: Anxiety and the effects of sodium lactate assessed clinically and physiologically. Br J Psychiatry 119:129–141, 1971

Klein DF: Delineation of two drug-responsive anxiety syndromes. Psychopharmacologia 5:397–408, 1968

Ko GN, Elsworth JD, Roth RH, et al: Panic-induced elevation of plasma MHPG levels in phobic-anxious patients: effects of clonidine and yohimbine. Arch Gen Psychiatry 40:425–430, 1983

Kopin IJ, Gordon EK, Jimerson DC, Polinsky RJ: Relation between plasma and cerebrospinal fluid levels of 3-methoxy-4-hydroxyphenylglycol. Science 219:73–75, 1983

Leusen I, Weyne J: Metabolic processes in the brain during respiratory and non-respiratory alkalosis and acidosis, in Acid Base Homeostasis of the Brain Extracellular Fluid and the Respiratory Control System. Edited by Loeschcke HH. Littleton, Mass, Thieme Edition/Publishing Sciences Group, 1975

Magarian GJ: Hyperventilation syndromes: infrequently recognized common expressions of anxiety and stress. Medicine (Baltimore) 61:219–236, 1982

Martin JB, Reichlin S, Brown GM (eds): Clinical Neuroendocrinology. Philadelphia, FA Davis Co, 1977

Nemoto EM, Severinghaus JW: Stereospecific permeability of rat blood-brain barrier to lactic acid. Stroke 5:81–84, 1974

Nemoto EM, Hoff JT, Severinghaus JW: Lactate uptake and metabolism by brain during hyperlactatemia and hypoglycemia. Stroke 5:48–53, 1974

Ninan PT, Insel TM, Cohen RM, et al: Benzodiazepine receptor-mediated experimental "anxiety" in primates. Science 218:1332–1334, 1982

Oldendorf WH: Blood brain permeability to lactate. Eur Neurol 6:49–55, 1972

Pauls DL, Bucher KD, Crowe RR, et al: A genetic study of panic disorder pedigrees. Am J Hum Genet 32:639–644, 1980

Pitts FN Jr, McClure JN: Lactate metabolism in anxiety neurosis. N Engl J Med 277:1329–1336, 1967

Redmond DE Jr, Huang YH: New evidence for a locus coeruleus-norepinephrine connection with anxiety. Life Sci 25:2149–2162, 1979

St. John WM: Integration of peripheral and central chemoreceptor stimuli by pontine and medullary respiratory centers. Fed Proc 36:2421–2427, 1977

Sheehan DV: Panic attacks and phobias. N Engl J Med 307:156–158, 1982

Sheehan DV, Ballenger J, Jacobsen G: Treatment of endogenous anxiety with phobic, hysterical, and hypochondriacal symptoms. Arch Gen Psychiatry 37:51–59, 1980

Sheehan DV, Sheehan KE, Minichiello WE: Age of onset of phobic disorders: a reevaluation. Compr Psychiatry 22:544–553, 1981

Skatrud JB, Dempsey JA, Kaiser DG: Ventilatory response to medroxyprogesterone acetate in normal subjects: time course and mechanism. J Appl Physiol 44:939–944, 1978

Weyne J, Leusen I: Lactate in CSF in relation to brain and blood, in Fluid Environment of the Brain. Edited by Cserr HF, Fenstermacher JD, Fencl V. New York, Academic Press, 1974